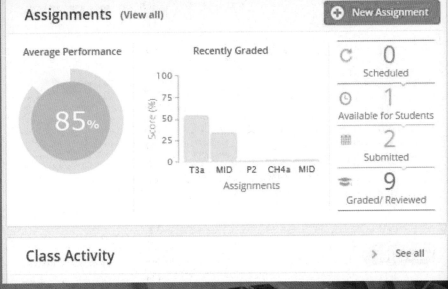

Professional Food Manager | 5th Edition

NEHA

NATIONAL ENVIRONMENTAL HEALTH ASSOCIATION

Updated to the 2015
supplement to the
2013 FDA Food Code

EDUCATION & TRAINING

WILEY

This book was set in 9/12 Helvetica Neue. Cover design and page layout by Wendy Lai. Printed and bound by Quad Versailles. The cover was printed by Quad Versailles.

Author: Christine Hollenbeck, CP-FS

Editor: Nancy Finney, MPA

This book is printed on acid-free paper. ∞

For general information on our other products and services, or technical support, please contact our Customer Care Department within the United States at 800-762-2974, outside the United States at 317-572-3993 or fax 317-572-4002.

Wiley also publishes its books in a variety of electronic formats. Some content that appears in print may not be available in electronic books.

For more information about Wiley products, visit our website at www.wiley.com.

Library of Congress Cataloging in Publication Data

Title: Professional food manager / National Environmental Health Association,
    NEHA.
Other titles: NEHA certified professional food manager.
Description: 5th edition. | Hoboken, New Jersey : John Wiley & Sons, Inc.,
    [2017] | Previous edition: NEHA certified professional food manager. |
    Includes bibliographical references and index.
Identifiers: LCCN 2016028275 (print) | LCCN 2016029033 (ebook) | ISBN
    9781119148524 (paperback : acid-free paper) | ISBN 9781119195054 (pdf) |
    ISBN 9781119195108 (epub)
Subjects: LCSH: Food industry and trade—Sanitation—United States. | Food
    handling—United States. | Food industry and trade—United
    States—Employees. | BISAC: TECHNOLOGY & ENGINEERING / Food Science.
Classification: LCC TP373.6 .N44 2017 (print) | LCC TP373.6 (ebook) | DDC
    363.19/20973—dc23
LC record available at https://lccn.loc.gov/2016028275

ISBN: 978-1-119-19510-8

Printed in the United States of America

10 9 8 7 6 5 4 3 2 1

CONTENTS

PREFACE   IX

CHAPTER ONE
INTRODUCTION TO FOOD SAFETY 1

Lesson 1 Safe Food 2

Lesson 2 Why Food Safety? 3

Lesson 3 Who Protects Our Food? 4

Assessment Questions 7

Questions for Discussion 8

CHAPTER TWO
FOODBORNE ILLNESS 9

Lesson 1 Foodborne Illness vs. Foodborne
          Illness Outbreak 10

Lesson 2 High-Risk Populations 12

Assessment Questions 15

Questions for Discussion 16

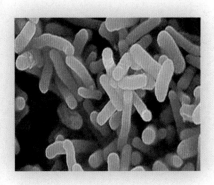

CHAPTER THREE
CONTAMINATION 17

Lesson 1 Contamination 18

Lesson 2 Bacteria 23

Lesson 3 Viruses 29

Lesson 4 Parasites and Fungi 32

v

Lesson 5 Chemical Contamination 35

Lesson 6 Natural Toxins 37

Lesson 7 Allergens 40

Assessment Questions 44

Questions for Discussion 45

CHAPTER FOUR
PEST CONTROL 46

Lesson 1 Pests 47

Lesson 2 Integrated Pest Management 49

Lesson 3 Pest Prevention 52

Lesson 4 Pesticides 56

Assessment Questions 59

Questions for Discussion 60

CHAPTER FIVE
EMPLOYEE TRAINING 61

Lesson 1 Hygiene 62

Lesson 2 Hand Washing 65

Lesson 3 Gloves 68

Lesson 4 Employee Health 69

Lesson 5 Communication 72

Lesson 6 Delivering Training 73

Assessment Questions 75

Questions for Discussion 76

CHAPTER SIX
FACILITIES AND EQUIPMENT 77

Lesson 1 Facility Design 78

Lesson 2 Food Contact Materials 82

Lesson 3 Cleaning and Sanitizing 86

Lesson 4 Washing Facilities 91

Lesson 5 Plumbing 95

Assessment Questions 98

Questions for Discussion 99

CHAPTER SEVEN
PURCHASING AND STORING FOOD 100

Lesson 1 Purchasing 101

Lesson 2 Suppliers: Transportation
and Delivery 102

Lesson 3 Storage 104

Assessment Questions 111

Questions for Discussion 112

CHAPTER EIGHT
SAFE FOOD HANDLING 113

Lesson 1 Time and Temperature 114

Lesson 2 Preparation 117

Lesson 3 Cooking 120

Lesson 4 Cooling and Reheating 123

Lesson 5 Service 125

Assessment Questions 127

Questions for Discussion 128

CHAPTER NINE
THE HACCP APPROACH
TO FOOD SAFETY 129

Lesson 1 HACCP Overview 130

Lesson 2 HACCP Principles 132

Assessment Questions 139

Questions for Discussion 140

CHAPTER TEN
FOOD SAFETY STANDARDS 141

Lesson 1 FDA Food Code 142

Lesson 2 Inspections 143

Lesson 3 Sampling 148

Lesson 4 Labeling 150

Lesson 5 Hazardous Materials 151

Assessment Questions 153

Questions for Discussion 154

Chapter Assessment Answer Key 155

Glossary 157

Index 161

PREFACE

Safe food handling and protection are critical components of any retail food service operation. Successful food managers understand that most foodborne illness is the result of poor personal hygiene, cross-contamination, or temperature abuse. Understanding these and other food safety risks — and how to prevent them — is the first step food managers can take to create and maintain a healthy environment for customers and employees. Food managers must work each day to set high standards and expectations for food safety, while continually adapting to and applying evolving industry guidelines within their facilities.

The National Environmental Health Association (NEHA) is the leader in environmental health education and protection. NEHA's *Professional Food Manager, Fifth Edition* provides the information food managers need to implement and maintain safe and healthful food practices. *Professional Food Manager, Fifth Edition,* includes the most up-to-date food safety trends, key principles of food safety management, health department guidelines, temperature recommendations, and basic sanitation procedures. Whether you are seeking a food manager certification or are already a food service professional, knowing and applying the best practices addressed in this book will help you develop an effective food safety program that can prevent foodborne illnesses.

WHAT'S NEW FOR THE FIFTH EDITION

Professional Food Manager Fifth Edition continues to offer streamlined, easy-to-understand information for increasing and maintaining a safe food service operation. It includes key information on topics such as managing suppliers, monitoring food temperatures, communication, and foodborne illness statistics.

In this edition, you'll also find:

- Current information, updated to the 2015 supplement of the 2013 U.S. Food and Drug Administration Food Code
- Latest information on reportable diseases
- New key terms and updated terminology, reflecting the 2015 supplement
- Reorganization of the topics for better flow of information
- Tables and charts placed throughout the text to be used for ongoing reference
- Discussion questions at the end of each chapter to engage real-life application of the principles presented in the text

SUPPLEMENTARY MATERIALS

In addition to this book, qualified instructors or trainers of the NEHA materials have access to a wide array of resource materials to support their classes through an easy-to-navigate companion website at www.wiley.com/college/neha, including:

- Instructor's Manual/Activity Guide, which contains chapter quizzes, true/false questions, matching exercises, fill-in-the-blank questions and more
- PowerPoint® Presentation with Discussion Notes
- Course Pre-Test and Answer Key
- Practice Examination and Answer Key
- Class Syllabi for 8-hour and 16-hour classes

- Case Studies with Discussion Questions
- Test Bank, specifically formatted for Respondus, an easy-to-use software program for creating and managing exams that can be easily printed to paper or published directly to Blackboard, WebCT, Desire2Learn, eCollege, ANGEL and other eLearning systems.

ABOUT THE NATIONAL ENVIRONMENTAL HEALTH ASSOCIATION

Headquartered in Denver, Colorado, the National Environmental Health Association (NEHA) is an educational and professional organization that supports and represents those who work in the environmental health and protection field. Founded in 1937, and now with 53 affiliated associations worldwide, NEHA's mission, "to advance the environmental health and protection professional for the purpose of providing a healthful environment for all" is as relevant today as it was when the organization was founded. NEHA works closely with state and local health departments, federal regulators, and the retail food industry to create and deliver the most effective, up-to-date food safety training available. All NEHA food safety programs are focused on providing industry professionals with the right training, certifications, and credentials they need to meet their education goals and advance their careers.

INTRODUCTION TO FOOD SAFETY

M any agencies work together to ensure that the U.S. food supply is safe. In the United States, food can travel long distances from the farm to the production facility, then to the distributor, and finally to the retail facility or restaurant. Occasionally, a breakdown in the system can result. One such example occurred in 2008, when a widely publicized *Salmonella* outbreak from peanut butter paste sickened 714 people, hospitalized 171, and killed nine.[1] The Peanut Corporation of America was found to be responsible for shipping the tainted product. More education, training, and understanding of the dangers associated with contaminated foods could potentially have prevented that tragedy. In retail environments, learning about food safety includes understanding all of the possible ways someone could become sick from eating food, as well as knowing all of the things food managers can do to prevent foodborne illness or injury—from the time the food is purchased to the time it enters the food facility until it is finally served to the customer.

After reading this chapter, you should be able to:

- Define food safety.
- Explain why learning about food safety is essential to your job.
- Describe food safety roles at the local, state, and federal levels.

1

It is important to practice safety in food preparation.
Edwin Remsberg/VWPics/Newscom

Learning Objective: Define food safety.

Food safety is a scientific discipline that describes the handling, preparation, and storage of food in ways that prevent foodborne illness. This includes the practices that food service employees should follow to avoid the potentially severe health hazards associated with unsafe food.

Safe food is free of contaminants and should not cause harm to the person consuming it.

Food can be kept safe by managing how that food is:

- Packaged
- Delivered
- Stored
- Prepared
- Cooked
- Served

The food manager's goal should always be to serve safe, wholesome food to every consumer who comes to the facility.

A consumer can become sick or injured when proper procedures are not followed. Food can hurt people in a number of different ways including: biting into something sharp within the food, eating food containing illness-causing bacteria, or ingesting poison left on the food as residue from cleaning supplies. Improperly cooked food is one of the main causes of foodborne illness.

Any time a food is considered unsafe, unwholesome, or impure, it is considered **adulterated food**. Food might become adulterated accidentally, unintentionally, or purposefully. If a food contains undisclosed ingredients that might cause an allergic reaction in certain individuals, it is also regarded as adulterated. False or misleading information on a label is called misbranding.

Further examples that cause food to be considered adulterated include:

- The food contains poisonous or unsanitary ingredients or anything that might cause injury or illness upon consumption.
- The food preparation, packing, or storing occurred under unsanitary or unsafe conditions.
- The food contains an ingredient not specifically labeled (such as an addition or substitution).

The food manager needs to understand: (1) what the term *adulterated food* means; (2) how a food might become adulterated; (3) what needs to be done once a food becomes adulterated; and (4) how to keep food from becoming adulterated in the first place.

LESSON 2 | WHY FOOD SAFETY?

Learning Objective: Explain why learning about food safety is essential to your job.

It is the position of the Academy of Nutrition and Dietetics that all people should have access to a safe food and water supply.[2] The Academy supports science-based food regulations and recommendations that are applied consistently across all foods and water regulated by all agencies, and incorporates traceability and recalls to limit food- and water-borne outbreaks.

The 2011 **Food Safety and Modernization Act (FSMA)** is the most sweeping reform to the U.S. food safety system in more than 70 years. FSMA aims to ensure that the U.S. food supply is safe by shifting regulators' focus from responding to contamination to preventing it.

When people purchase food in a restaurant or grocery store, they have confidence that eating the food will not make them sick or die. Yet, each year:

- One in six Americans, or 48 million people, gets sick from contaminated foods.
- 3,000 people die from contaminated foods.
- $365 million is spent in direct medical costs on illnesses from *Salmonella* bacteria alone.
- Reducing foodborne illnesses by just 10 percent would keep nearly five million Americans from getting sick every year.[3]

By learning about the causes and prevention of foodborne illness and following the food safety processes outlined in this text, many lives can be saved.

There are additional benefits to practicing good food safety in the food facility:

- Satisfied customers
- Good reputation
- Increased business
- Legal compliance
- Minimal food waste
- Good working conditions
- Higher staff morale
- Reduced staff turnover
- Increased productivity
- Better relationship with enforcement officer
- Higher profits

Likewise, there are consequences to poor food safety procedures:

- Foodborne illness outbreaks
- Food contamination
- Customer complaints
- Pest infestations
- Food waste
- Closure of premises
- Fines and civil action
- Lower profits
- Loss of business

Reducing foodborne illness by 10% would keep almost 5 million Americans from getting sick each year.

Caia Images/SuperStock

Strict food temperature controls, protection from contamination, and good personal hygiene procedures will prevent most instances of foodborne illness. Knowing who to contact regarding specific food safety questions, and when, can help a food safety manager and the entire food facility staff stop foodborne illness from occurring or spreading.

Learning Objective: Describe food safety roles at the local, state, and federal levels.

Government agencies are responsible for setting food safety standards, conducting inspections, and ensuring that standards are met. These agencies also maintain strong enforcement programs to deal with noncompliance.

REGULATORY AUTHORITIES

The regulatory authority involved with a food facility varies based on state regulations. Food managers should be familiar with their local governing bodies and with how these agencies regulate or enforce food safety.

The levels of government that are involved in food safety regulation and enforcement include:

- Local (for example, cities, towns, and others)
- County
- State
- Tribal
- Territorial
- Federal

Local, county, tribal, territorial agencies

Local and county agencies typically have a closer connection with the retail food facilities. Local and state health departments provide food safety inspections and work with food managers on compliance. These agencies are important in emergencies, such as a fire, flood, or foodborne illness outbreak. Food managers should have the contact information for these local agencies readily available and in an accessible location in the event that employees or management must quickly reach an agency. Enforcement methods and systems vary by jurisdiction, and managers should become familiar with their agency and the regulations the facility must follow.

State agencies

State agencies enforce laws and adopt regulations to ensure that local agencies are fulfilling their duties. Each state determines how much of the FDA Food Code to adopt. Some states adopt the latest updated Food Code, some use an older version, and others only adopt portions of the Food Code. Some states are strict, while others are lenient with their regulations. If any issues arise with a local authority, then the food manager should contact the appropriate state agency.

Federal agencies

The federal government is involved with food safety throughout the entire food chain. Federal agencies are dedicated to ensuring safe food for consumers in the United States. The following agencies all play a role in the food safety process in one way or another.

The U.S. Food and Drug Administration (FDA):

- Assures that certain foods are safe, wholesome, sanitary, and properly labeled
- Regulates all foods and food ingredients introduced into or offered for sale in interstate commerce except for those regulated through the **USDA**
- Keeps a registry of facilities that manufacture, process, pack, or hold food that is intended for human or animal consumption in the United States

Inspections enforce safety standards.

Phil Boorman Cultural/Newscom

- Provides information about the regulatory process and develops the Food Code in an effort to educate the industry and protect consumers
- Works with the Center for Food Safety and Applied Nutrition (CFSAN) in the field for the safety of **food additives** and ingredients, including radiation and other food contaminants, labeling, and consumer education

The U.S. Department of Agriculture (USDA):

- Provides leadership on food, agriculture, natural resources, rural development, nutrition, and related issues based on public policy, the best available science, and effective management
- Includes the Food Safety and Inspection Service (FSIS), which regulates aspects of the safety and labeling of domesticated animals, such as cattle and chicken, as well as certain egg products

The Centers for Disease Control and Prevention (CDC):

- Provides the vital link between illness in people and the food safety systems of government agencies and food producers
- Helps to determine any possible links to foodborne illness outbreaks and confirm the possible cause

Food safety depends on strong partnerships. The **CDC**, **FDA**, and FSIS all collaborate at the federal level to promote food safety.

The National Oceanic and Atmospheric Administration (NOAA):

- Runs **NOAA** Fisheries (also known as the National Marine Fisheries Service, or NMFS)
- Is responsible for inspecting fish, and fish farming and harvesting practices, to ensure safe sources of seafood as well as sustainable, protected, healthy fish and underwater ecosystems

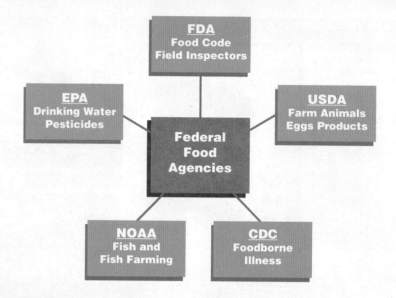

The Environmental Protection Agency (EPA):

The Environmental Protection Agency (**EPA**) is responsible for protecting human health and the health of the environment. The EPA:

- Regulates contaminants in drinking water and ensures proper water treatment

- Develops national standards for drinking water from municipal water supplies (tap water) to limit the levels of impurities

- Regulates many aspects of pesticides, including setting limits on how much of a pesticide may be used on food during growing and processing and how much can remain on food for consumption

KEY TERMS

Adulterated food   Food that is generally impure, unsafe, or unwholesome.

CDC   Centers for Disease Control and Prevention.

EPA   Environmental Protection Agency.

FDA   Food and Drug Administration.

Food additives   Preservatives, food colorings, and flavorings that are added to food.

Food Safety Modernization Act (FSMA)
The act that aims to ensure the U.S. food supply is safe by shifting the focus from responding to contamination to preventing it.

Food safety   A scientific discipline that describes the handling, preparation, and storage of food in ways that prevent foodborne illness.

NOAA   National Oceanic and Atmospheric Administration (NOAA Fisheries is also known as the National Marine Fisheries Service, or NMFS).

Safe food   Food that is free of contaminants and does not cause harm to the person consuming it.

USDA   United States Department of Agriculture.

REFERENCES

[1] Multistate outbreak of *Salmonella* Typhimurium infections linked to peanut butter. (2009, May 11). Retrieved from http://www.cdc.gov/salmonella/2009/peanut-butter-2008-2009.html

[2] Position of the Academy of Nutrition and Dietetics: food and water safety. (2014, October 24). Retrieved from http://www.ncbi.nlm.nih.gov/pubmed/25439082

[3] CDC and food safety. (2015, August 31). Retrieved from http://www.cdc.gov/foodsafety/cdc-and-food-safety.html

CHAPTER ONE

ASSESSMENT QUESTIONS

1. Which of the following is NOT a part of food safety?
   a. Food handling
   b. Preparation
   c. Storage
   d. Profits

2. Safe food is:
   a. Food that is not likely to harm the person consuming it
   b. Food that is low in calories
   c. Food that is low in sodium
   d. Food that is prepackaged or freeze-dried

3. The agency that deals with cattle processing is the:
   a. FDA
   b. USDA
   c. CDC
   d. EPA

4. The agency to call first when a water pipe bursts inside the kitchen of a food facility is:
   a. The local health department
   b. The state health department
   c. The federal health department
   d. The EPA

5. Each of the following represents a significant method of controlling foodborne illness except for:
   a. Increased productivity
   b. Temperature controls
   c. Protection from contamination
   d. Good personal hygiene

6. The term *adulterated food* means food that is:
   a. Unsanitary
   b. Stored under unsafe conditions
   c. Made with an unidentified ingredient
   d. All of the above

7. If a customer returns to your facility complaining of food poisoning that he or she believes was contracted from your facility, the correct response would be to:
   a. Call the local health department
   b. Contact the FDA
   c. File a foodborne illness report with the state health department
   d. Notify the staff and put up posters warning customers to keep an eye out for illness

8. The number-one reason for taking a food safety course is to:
   a. Acquire a food safety manager certification
   b. Ensure the public does not contract foodborne illness
   c. Pass the health inspection
   d. Keep staff and customers happy

9. Which federal agency develops national standards for drinking water?
   a. NOAA
   b. EPA
   c. CDC
   d. DFA

10. The FSIS regulates aspects of the safety and labeling of:
    a. Marine life like fish and shellfish
    b. Peanuts, tree nuts, and dairy products
    c. Toxic chemicals found in cleaning products
    d. Domesticated animals, like cattle and chickens

QUESTIONS FOR DISCUSSION

1. Which agency should be contacted during a power outage in a food facility?
2. Which agency is responsible for inspecting animal feed production facilities?
3. In what ways is the CDC involved in foodborne illness outbreaks?

FOODBORNE ILLNESS

The food supply in the United States is among the safest in the world. However, when certain disease-causing bacteria or other pathogens contaminate food, they can cause foodborne illness, often called food poisoning. Foodborne illness continues to be an important problem in the United States. In 2013, California laboratory findings linked an outbreak of *Salmonella* infections to Foster Farms brand chicken.[1]

- 634 people were infected with *Salmonella*.
- 29 states and Puerto Rico were affected.
- The outbreak occurred over a four-month period.
- 38 percent of ill people were hospitalized.
- 77 percent of those who were ill were from one area in California.

- 74 percent of ill people reported consuming chicken made at home the week before becoming sick.
- 87 percent of the people reported consuming Foster Farms chicken.
- A cluster of 25 ill people had purchased cooked rotisserie Foster Farms chickens from the same store during the outbreak.

Foster Farms initiated a recall of their chicken and chicken products that could have been contaminated with *Salmonella*, and the instances of foodborne illnesses associated with the chicken decreased. Luckily, no one died as a result of the incident; however, the number of people who were sickened by the product is startling. As is the case with most foodborne illnesses, this was entirely preventable.

After reading this chapter, you should be able to:

- Identify ways that a person could contract a foodborne illness.

- Discuss high-risk populations and best practices for protecting them.

Common symptoms of a foodborne illness are abdominal pain, diarrhea, vomiting, and nausea.
vadimguzhva/iStock

| # FOODBORNE ILLNESS VS. FOODBORNE ILLNESS OUTBREAK

Learning Objective: Identify ways that a person could contract a foodborne illness.

A **foodborne illness** is defined as an infection or intoxication that results from consuming foods contaminated with harmful microorganisms or toxins.

Symptoms of foodborne illnesses depend on the cause. Symptoms can range from mild to serious and can last from a few hours to several days.

Common symptoms of many foodborne illnesses include:

- Vomiting
- Diarrhea or bloody diarrhea
- Abdominal pain
- Fever
- Chills

Some bacteria and **toxins** produced by bacteria affect the nervous system, causing other symptoms, such as:

- Headache
- Tingling or numbness of the skin
- Blurred vision
- Weakness
- Dizziness
- Paralysis

Harmful bacteria are the most common cause of foodborne illness, but there are many other causes, including:

- Viruses
- Fungi (molds and yeasts)
- Parasites
- Natural toxins
- Allergens

Serious long-term effects associated with several common types of foodborne illness include:

- Kidney failure
- Chronic arthritis
- Brain and nerve damage
- Death

When two or more cases of foodborne illness occur during a limited period of time with the same organism and are associated with either the same food service operation, such as a restaurant, or the same food product, it is called a **foodborne illness outbreak**.

There is, however, one exception to the "two or more cases of foodborne illness" rule: botulism. Because of the severity of the illness and the possibility that a source food may cause others to become seriously ill, a single case of botulism is considered an outbreak.

The Centers for Disease Control and Prevention (CDC) have identified five risk factors that lead to the majority of foodborne illnesses:

- Improper hot/cold holding temperatures of potentially hazardous food
- Improper cooking temperatures of food
- Dirty and/or contaminated utensils and equipment
- Poor employee health and hygiene
- Food from unsafe sources

Each of these risk factors is tied directly to harmful microorganisms, such as bacteria.

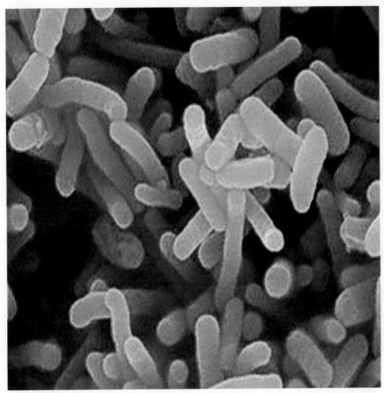

Bacteria
Sebastian Kaulitzki, 2014/Shutterstock

FOODBORNE ILLNESS OUTBREAKS

Foodborne illness outbreaks appear to be decreasing. Part of the reason is increased surveillance and better detection methods; public health agencies can identify and link sporadic cases from different states or regions that are caused by the same organism.

Despite improvements in identifying outbreaks, new disease-causing organisms have emerged that are resistant to the old methods of destroying them. Automation in food production and far-reaching, rapid distribution mean that the ways foodborne illness can arise and spread are changing, and the scope of outbreaks can be much larger than before. Multistate and multinational occurrences are not uncommon.

Local health agencies inspect food service and food retail facilities, provide technical assistance to **food establishments**, and educate consumers about food safety. Local agencies are the front line in protecting the public from foodborne illness associated with improper handling of food or poor hygienic practices in food service and food retail operations. The food protection manager needs to stay aware of any current issues in food safety and know how to contact the local food inspector should any questions about the safety of food arise.

CDC risk factors include improper cooking temperatures of food.
James "BO" Insogna/Shutterstock

Learning Objective: Discuss high-risk populations and best practices for protecting them.

One in six Americans will get sick from food poisoning this year.[2] Most of them will recover without any lasting effects from their illness. For some, however, the effects can be devastating and even deadly.

Those at greater risk, known as **high-risk populations**, include:

- Infants
- Young children
- Pregnant women and their unborn babies
- The elderly
- People with weakened immune systems (**immunocompromised**.)

Some people may become ill after ingesting only a few harmful bacteria; others may remain symptom-free after ingesting thousands.

INFANTS AND YOUNG CHILDREN

Infants and young children are more at risk for foodborne illness because their immune systems are still developing.

PREGNANT WOMEN AND THEIR UNBORN BABIES

Changes during pregnancy alter the mother's immune system, making pregnant women more susceptible to foodborne illness. Harmful bacteria can also cross the placenta and infect an unborn baby, whose immune system is underdeveloped and not able to fight infection. Foodborne illness during pregnancy is serious and can lead to miscarriage, premature delivery, stillbirth, sickness, or the death of a newborn baby.

OLDER ADULTS

As people age, their immune system and other organs become sluggish in recognizing and ridding the body of harmful bacteria and other pathogens that cause infections, such as foodborne illness. Many older adults have been diagnosed with one or more chronic conditions, such as diabetes, arthritis, cancer, or cardiovascular disease, and are taking at least one medication. The chronic disease process and/or the side effects of some medications may also weaken the immune system. In addition, stomach acid decreases as people get older, and this is significant because stomach acid plays an important role in reducing the number of harmful bacteria in the intestinal tract. Increased bacteria levels can increase the risk of illness.

PEOPLE WITH WEAKENED IMMUNE SYSTEMS

The immune system is the body's natural reaction or response to "foreign invasion." In healthy people, a properly functioning immune system readily fights off harmful bacteria and other pathogens that cause infection. However, the immune systems of transplant patients and people with certain illnesses, such as HIV/AIDS, cancer, and diabetes, are often weakened from the disease process and/or the side effects of some treatments, making them susceptible to many types of infections—like those that can be brought on by harmful bacteria that cause foodborne illness. In addition, diabetes may lead to a slowing of the rate at which food passes through the stomach and intestines, allowing harmful foodborne pathogens an opportunity to multiply.

Monkey Business Images,
2014/Shutterstock

pinkcandy, 2014/Shutterstock

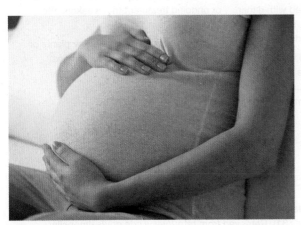

Africa Studio/Shutterstock

Sam Edwards/OJO Images/SuperStock

High-risk populations

If someone is in one of these high-risk groups, it's especially important to practice safe food handling. Vulnerable people are not only at increased risk of contracting a foodborne illness, but they are also more likely to have a lengthier illness, undergo hospitalization, or even die. If someone is at greater risk of foodborne illness, then he or she is advised not to eat:

- Raw or undercooked meat or poultry
- Raw fish, partially cooked seafood (such as shrimp and crab), and refrigerated smoked seafood
- Raw shellfish (including oysters, clams, mussels, and scallops) and their juices
- Unpasteurized (raw) milk and products made with raw milk, like yogurt and cheese
- Soft cheeses made from unpasteurized milk, such as feta, Brie, Camembert, blue-veined cheeses, and Mexican-style cheeses (such as such as queso fresco, queso panela, queso blanco, and asadero)
- Raw or undercooked eggs or foods containing raw or undercooked eggs, including certain homemade salad dressings (such as Caesar salad dressing), homemade cookie doughs and cake batters, and homemade eggnog

  NOTE: Most premade foods from grocery stores, such as Caesar dressing, premade cookie dough, or packaged eggnog, are made with pasteurized eggs, which are considered safer.

- Unwashed fresh vegetables, including lettuce and salads
- Unpasteurized fruit or vegetable juices (these juices carry a warning label)

- Hot dogs, luncheon meats (cold cuts), fermented and dry sausages, and other deli-style meats, poultry products, and smoked fish—unless they are reheated until steaming hot
- Deli salads (without added preservatives) prepared on-site in a deli-type establishment, such as ham salad, chicken salad, or seafood salad
- Unpasteurized, refrigerated pâtés or meat spreads
- Raw sprouts (alfalfa, bean, or any other sprout)

KEY TERMS

Food establishment   Any business whose commercial operations deal with food or food sources.

Foodborne illness   An infection or intoxication that results from consuming foods contaminated with harmful microorganisms or toxins.

Foodborne illness outbreak   An instance when two or more cases of foodborne illness occur during a limited period of time with the same organism and are associated with either the same food service operation, such as a restaurant, or the same food product.

High-risk populations   People who are more likely to contract foodborne illness, including the elderly, the very young, people who are immunocompromised, and pregnant women.

Immunocompromised   Having an immune system that is impaired, including the very old, the very young, and those with a disease or ongoing treatment that weakens the immune system.

Toxin   A substance created by plants or animals that is poisonous to humans.

REFERENCES

[1] Multistate outbreak of multidrug-resistant *Salmonella* Heidelberg infections linked to Foster Farms brand chicken (final update). (2014, July 31). Retrieved from http://www.cdc.gov/salmonella/heidelberg-10-13/

[2] Estimates of foodborne illness in the United States. (2014, January 8). Retrieved from http://www.cdc.gov/foodborneburden/

ASSESSMENT QUESTIONS

1. Which of the following lunch choices would be safest for a pregnant woman to eat?
   a. Hot dog
   b. Yogurt parfait
   c. Seared ahi tuna salad
   d. Chicken sandwich

2. Which of the following salad bar selections is least likely to contribute to foodborne illness?
   a. Crumbled feta cheese
   b. Raw sprouts
   c. Raisins
   d. Crab meat

3. Common symptoms of many foodborne illnesses include each of the following EXCEPT:
   a. Diarrhea
   b. Watery eyes
   c. Cramping
   d. Fever and chills

4. The most common cause of foodborne illness is:
   a. Viruses
   b. Parasites
   c. Allergens
   d. Bacteria

5. Soufflé made for elderly persons must use what type of eggs?
   a. Hard-boiled
   b. Free-range
   c. Pasteurized
   d. Organic

6. Which of the following is NOT one of the five risk factors that lead to the majority of foodborne illnesses, according to the CDC?
   a. Improper cooking temperatures of food
   b. Poor employee health and hygiene
   c. Bacteria in and around the food environment
   d. Dirty and/or contaminated utensils and equipment

7. A single case of which of the following illnesses is considered a foodborne illness outbreak?
   a. Botulism
   b. Hepatitis A
   c. Diabetes
   d. *Salmonella*

8. Which of the following is NOT an example of a high-risk population?
   a. Elderly people in a nursing home
   b. Pregnant women
   c. Infants in a nursery
   d. Teenagers in a large high school

9. Which of the following food items is least likely to contribute to foodborne illness?
   a. Oysters
   b. Pasteurized milk
   c. Alfalfa sprouts
   d. Cake batter

10. Why are infants and young children more at risk for foodborne illness?
    a. Their immune systems are still developing.
    b. They may have low birth weights.
    c. Their parents are still learning what foods to feed them.
    d. They are not considered a high-risk population.

QUESTIONS FOR DISCUSSION

1. List some high-risk populations. Why are they considered high-risk? How can you as a food handler protect them from foodborne illness in your establishment?

2. Imagine you are the new manager of a restaurant with a popular salad bar. What are some items to exclude from the salad bar in an effort to reduce the risk of foodborne illness in your restaurant?

CHAPTER THREE

CONTAMINATION

Foodborne illness can be contracted in many ways, including by ingesting food or water that has been contaminated by an ill food service worker. Unfortunately, it's an all-too-common scenario: After a nice night at a popular restaurant, a consumer wakes up the following day feeling extremely ill, and is diagnosed with a foodborne illness, such as norovirus—the most common cause of acute **viral gastroenteritis** in the United States.[1] The successful food safety manager can help keep scenarios like this from happening in the first place by learning how to prevent contamination throughout the food facility and by implementing practices that keep food safe.

After reading this chapter, you should be able to:

- Explain some of the ways in which food can become contaminated.
- List bacteria that can cause foodborne illness.
- Describe the characteristics of viruses.
- Describe the characteristics of parasites and fungi.

- Identify best practices for preventing chemical contamination.
- Identify the natural toxins that can cause foodborne illness.
- List the major food allergens.

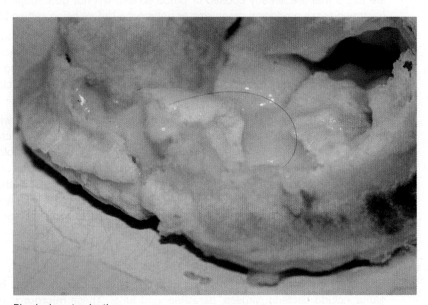

Physical contamination

Learning Objective: Explain some of the ways in which food can become contaminated.

Contamination is the presence of physical, chemical, or biological matter in or on food or the food environment. Food may be contaminated before delivery to a business, as a result of poor hygiene practices, or by customers if the food is not adequately protected.

Cross-contamination is one of the most common and dangerous types of contamination. Cross-contamination occurs when bacteria from contaminated foods (usually raw) transfer to other foods. Cross-contamination can happen in a variety of ways:

- **Directly**, e.g., when raw meat touches another food
- **By drip**, e.g., if raw meat incorrectly stored above ready-to-eat food and drips blood onto that food
- **Indirectly**, e.g., if raw food and ready-to-eat food are prepared with the same equipment

Because food is susceptible to cross-contamination at any point from farm to fork, it is vital to have controls in place to prevent adulteration. For example, using separate preparation areas to prepare raw poultry, meat, or fish than are used for already cooked or prepared foods is a simple way to prevent cross-contamination. If separate prep areas are not possible, prepare foods at different times and be sure to clean and sanitize the area before moving from raw meat, poultry, or fish preparation to ready-to-eat food preparation. **Ready-to-eat (RTE) foods** are foods that are already cooked or prepared and will not go through further treatment to destroy disease producing organisms or pathogens before being served to the consumer.

It is extremely important that food handlers are familiar with proper food storage practices. For example, raw foods such as eggs, meat, and poultry must always be placed below ready-to-eat foods, such as lettuce and tomatoes. Frozen, commercially processed and packaged raw animal foods may be stored or displayed with or above frozen, commercially processed and packaged, ready-to-eat foods.

Never use the same equipment for handling raw and ready-to-eat foods without cleaning and sanitizing in between. Use a color-coding system to reduce the risk of cross-contamination by ensuring the same equipment is not used for different food types. Color coding does not eliminate the need to clean and sanitize equipment. Color coding can be used for many types of equipment, such as cutting boards, knife handles, work surfaces, cloths, protective clothing, and packaging materials. Systems of color coding may differ from one establishment to another, so it is important that staff know and understand the color codes in their workplace.

BIOLOGICAL CONTAMINATION

Biological contamination is the most common type of contamination. Biological contaminants include bacteria, viruses, parasites, and fungi. These **microorganisms**, too small to be seen without a microscope, may be transferred to food from a variety of sources, such as people, raw food, contact surfaces, pests, and refuse. Biological contamination usually occurs as a result of ignorance, inadequate space, or poor structural design. It may also occur when food handlers take shortcuts or do not implement good hygiene practices. Biological contamination can have very serious consequences; in the early stages, it may not be detectable, but it may result in food spoilage, food poisoning, or even death.

CHEMICAL CONTAMINATION

Chemical contamination is the presence of unwanted chemical components in food or the food environment. Chemical contamination may be as basic as pesticide residue from improperly cleaned produce or as obscure as toxic metal poisoning from improper food preparation or storage equipment.

PHYSICAL CONTAMINATION

Physical contamination occurs when any foreign object is in or on food and presents a hazard or nuisance to those consuming it. Foreign objects can range from a piece of hair to naturally occurring food objects such as bones or stalks. Examples of common foreign objects that cause physical contamination risk are:

- Dirt
- Hair, skin, scabs, and fingernails
- Pencils, pens, and ink
- Jewelry
- Glass, metal fragments, wood, and paint chips
- Paper fragments
- Plastic and other food-packaging items, such as twist ties and staples
- Dead insects, rodents, and rodent droppings

Physical contamination can occur at any stage in the food creation, delivery, or preparation process.

As a first step to preventing physical contamination, carefully inspect all food deliveries to ensure that no physical contamination has occurred prior to delivery. Warning signs that food has been compromised can include damaged packaging and food that is not properly protected or presented. In addition, carefully screen all suppliers to ensure their food production facilities meet appropriate standards. Don't hesitate to replace a supplier if they repeatedly deliver food in poor condition.

Once food is delivered, make sure that it is properly covered and stored at the location. Regularly inspect food storage areas for cleanliness and signs of pest problems. Also make sure storage areas are well maintained. For example, if a lightbulb breaks in a storage area, ensure that the glass is immediately cleaned up and all food in the area is inspected to verify that it hasn't been contaminated with glass. Shatterproof lightbulbs or shields covering exposed bulbs are the best bet and are required in food preparation areas.

Food is probably most vulnerable when it is being prepared for consumption. For example, food can be contaminated when:

- Employees are removing food from storage and placing it on food preparation workstations.
- Employees are touching food with their hands and exposing food to any contaminants they may have on their person, such as jewelry or hair.
- Food preparation equipment and utensils have not been adequately cleaned, sanitized, maintained, and replaced when damaged.

To minimize the likelihood that food will be contaminated during preparation, ensure that good food safety policies are in place and that employees are well trained in these policies. Examples of good food safety policies include:

- Cleaning and inspecting workstations prior to beginning food preparation
- Thorough hand washing prior to handling food
- Ensuring that all employees wear hairnets and, if applicable, beard nets
- Not allowing jewelry in the food preparation area
- Not allowing loose items in shirt pockets, such as pens, pencils, paper, or coins

Food preparation
Wasant/Shutterstock

Good maintenance practices can also minimize the possibility of accidental physical contamination. Regularly inspect utensils and other food preparation equipment. Replace those items that are failing or might otherwise present a contamination risk. In addition, ensure a prompt and thorough cleanup after any mechanical repairs. This reduces the risk that any loose parts or waste materials, such as screws or drill shavings, might pose a contamination threat.

If using food displays, customers themselves can pose a physical contamination risk. To minimize this risk, food managers must:

- Ensure that display cabinets are kept clean and provide a good barrier between the food and the customer.
- Inspect all food from the display case prior to delivering it to the customer.
- Prohibit customers or any other unauthorized persons from accessing food prior to it being displayed for consumers.

INTENTIONAL CONTAMINATION

It is an unfortunate reality that in today's world, food managers must also prevent the purposeful contamination of food. Intentional contamination of food can be a goal not just of activist groups and terrorist organizations, but also of employees, former employees, and competitors.

As with other forms of food contamination, intentional contamination can occur at any stage in the food creation, delivery, or preparation process. To prevent intentional contamination of food, food safety must be taken as seriously as the safety of money, people, and equipment. Adequate food safety requires that each food establishment have its own comprehensive approach. While it is challenging to set up policies, procedures, and training for a comprehensive food safety program, remember that a single instance of intentional contamination can be catastrophic for the business.

FOOD DEFENSE

A good food safety program can make it difficult for any intentional contamination to take place. To be comprehensive, the program must take into account the most common areas of food vulnerability. This includes the people element as well as the building element. In addition, when considering the building element, both interior and exterior vulnerabilities are important.

Some basic considerations for a food defense program include the following:

- External physical security measures
- Internal process control security measures
- Personnel security measures
- Product and supply security measures
- Crisis management response security measures
- Internal and external communication programs
- Maintenance of customer confidence

PEOPLE

Let's consider the human aspect of intentional contamination. The list below shows some common sense practices that food managers can implement to help prevent intentional contamination at a food facility.

- Never allow anyone but on-duty employees in food preparation and storage areas.
- Limit on-duty employees to carrying essential items only while at work.
- Consider a two-employee rule for food preparation areas.
- Carefully monitor food preparation areas.
- Carefully screen employees prior to their hire. While it can be a tedious part of the employment process, identity verification and reference checks can help to maintain a safe food and work environment.

It is just as important to carefully screen and approve suppliers as it is to screen employees. Best practices for ensuring supplier safety include:

- Ensuring that all approved suppliers are clearly identified
- Inspecting and documenting all deliveries, and never allowing any deliveries from non-authorized vendors
- Asking suppliers about their food safety programs and whether tamper-evident packaging is available

Poisoning food
Vladimir Mucibabic/Shutterstock

Locked back door
Smith1972/Shutterstock

BUILDING

The following describes some of the practices food managers can implement within the facility to prevent intentional contamination:

- Controlling the entrances and exits for employee-only areas, such as food preparation areas, food display areas, and kitchens
- Using good lighting and cameras to eliminate hiding areas in all parts of the building
- Ensuring that self-service areas such as salad bars are carefully monitored
- Ensuring that the outside of the establishment also meets good safety standards, including a well-lit exterior, locked back doors, and a ventilation system that can be accessed only by authorized personnel
- Taking steps to ensure that no unauthorized individual, even employees, can access the building after normal business hours

A good food safety program addresses vulnerabilities from people as well as basic facilities safety. However, the best food safety program is useless if employees are unaware of it or don't follow it. It is vital that all employees are thoroughly trained in food safety practices and are required to follow them. In addition, employees should be trained and encouraged to report any suspicious activity that may be a sign of intentional food tampering.

LESSON 2 | BACTERIA

Learning Objective: List bacteria that can cause foodborne illness.

Bacteria are small single-celled organisms found almost everywhere on Earth. Most bacteria in the human body are harmless and some are even helpful. There are a small number of bacteria that spoil food, referred to as **food spoilage bacteria**, and others that cause illness, known as **pathogens**, which include food-poisoning bacteria. Some species of bacteria can live in extreme temperature conditions and environments.

PATHOGENIC BACTERIA

Many types of bacteria cause foodborne illnesses. Examples include:

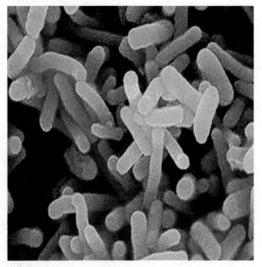

Bacteria
Sebastian Kaulitzki, 2014/Shutterstock

- ***Bacillus cereus*** (*B. cereus*), This type of bacteria produces two types of toxins: emetic and diarrheal. *B. cereus*, a bacterium that is found in starchy foods (**emetic** toxin), like rice and potatoes, as well as in meat products (diarrheal toxin). The emetic toxin affects a person within 30 minutes to six hours and causes nausea and vomiting. The diarrheal toxin illness starts in six to 15 hours and causes abdominal cramping and watery diarrhea.

- ***Campylobacter jejuni*** (*C. jejuni*), found in raw or undercooked chicken and raw milk; illness starts in two to five days and causes diarrhea and abdominal pain.

- ***Clostridium botulinum*** (*C. botulinum*), a bacterium that may contaminate improperly canned foods and smoked and salted fish. Illness starts in 18 to 36 hours and causes dizziness and double vision. A very small amount of *Clostridium botulinum* toxin can cause botulism, a deadly foodborne illness.

- ***Clostridium perfringens*** (*C. perfringens*), which is found in stews, casseroles, and meat pies. Illness starts in eight to 22 hours and causes diarrhea and abdominal pain.

- ***Escherichia coli*** (*E. coli*), which includes several different strains, only a few of which cause illness in humans. *E. coli* O157:H7 is the strain that causes the most severe illness. Common sources of *E. coli* include raw or undercooked hamburger, unpasteurized fruit juices and milk, and fresh produce. Illness starts in three to five days and causes severe abdominal pain and watery diarrhea.

- ***Listeria monocytogenes*** (*L. monocytogenes*), found in raw and undercooked meats, unpasteurized milk, soft cheeses, ready-to-eat deli meats, and hot dogs. Illness starts in several days to three weeks and causes a persistent fever and flu-like symptoms.

- ***Salmonella***, spp., a bacterium found in many foods, including raw and undercooked meat, poultry, dairy products, and seafood. *Salmonella* may also be present on eggshells and inside eggs. Illness from *Salmonella* starts in six to 48 hours and causes nausea and vomiting.

- ***Shigella***, spp., several bacteria found in deli salads, dairy and poultry; illness starts in 12 to 48 hours and causes diarrhea and blood in the stool.

- ***Staphylococcus aureus*** (*S. aureus*), which is found in dairy products, deli salads, and custards. Illness starts in 30 minutes and causes nausea and vomiting.

- ***Toxoplasma gondii*** (*T. gondii*), found in raw or undercooked meat and seafood or cat feces. Onset time is anywhere from five to 23 days and can last several weeks.

- ***Vibrio parahaemolyticus*** (*V. parahaemolyticus*), a bacterium that may contaminate fish or shellfish. Illness starts in four hours to four days and causes diarrhea, vomiting, fever, and chills.

Bacteria are found all over: on food, in water, in soil and air, and on or in animals and pests. Bacteria can be found on and in people, and not necessarily because they are ill. People who show no symptoms of illness, but excrete food-poisoning bacteria or simply carry them on their bodies, are referred to as **carriers** and are a source of food-poisoning bacteria.

Some harmful bacteria may already be present in foods when they are purchased. Raw foods, including meat, poultry, fish and shellfish, eggs, unpasteurized milk and dairy products, and fresh produce, often contain bacteria that cause foodborne illnesses. Bacteria can contaminate food—making it harmful to eat—at any time during growth, harvesting or slaughter, processing, storage, and shipping.

Foods may also be contaminated with bacteria during food preparation. If food preparers do not thoroughly wash their hands, kitchen utensils, cutting boards, and other kitchen surfaces that come into contact with raw foods, cross-contamination may occur.

STRUCTURE OF BACTERIA

Seen through a microscope, the shape of the bacteria can help to identify the type of bacteria. Bacteria vary considerably in shape:

- Cocci are spherical.
- Rods are sausage-shaped.
- Spirochetes are spiral.
- Vibrios are comma-shaped.

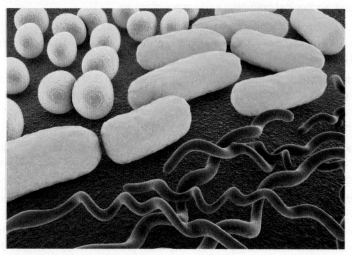

Cocci, Rods, and Spirochetes
Pasieka/Science Source

In simple terms, bacteria have a cell wall, which gives shape to them and holds together the cytoplasm and nuclear material inside. Flagella are attachments to the cell wall, which allow some bacteria to move in liquids. Some bacteria also have an outer slime layer, called a capsule. Large numbers of bacteria may cause visible effects on food—for example, a slime layer on the surface of spoiled meat.

BACTERIAL SPORES

Some bacteria produce **spores**, also known as **bacterial spores**, which enable them to survive high temperatures. The phase when spores are formed is known as sporulation. Spores are activated by cooking and **germinate** during slow cooling. Spores are very difficult to kill and will also protect bacteria from other adverse conditions, such as drying and sanitization. Spores can survive for many years, without the need for food or water, but they do not multiply. It is only when favorable conditions return that they start to multiply again. Bacteria that produce spores are:

- *Clostridium perfringens*
- *Clostridium botulinum*
- *Bacillus cereus*

CLASSIFYING BACTERIAL ILLNESS

Bacterial food poisoning may be **toxic** (directly poisonous) or **infectious** (spreading among people). Toxic bacteria cause intoxications and infectious bacteria cause infections.

Intoxication occurs when bacteria produce and release exotoxins directly into the food that a person eats. **Exotoxins** are poisons produced during the multiplication of those bacteria. They are highly toxic proteins and are often produced in food. Illnesses caused by intoxication normally have a short **onset time**; that means that the symptoms will come on quickly. Chemical residues and food additives can also cause intoxication.

Infection occurs when bacteria release endotoxins in the intestine of the affected person after that person has ingested the food. An **endotoxin** is a poison present in the cell wall of bacteria that is released upon the death of the bacteria. Illnesses caused by infection normally have a longer onset time; it may take one to two days before the infection makes a person feel ill.

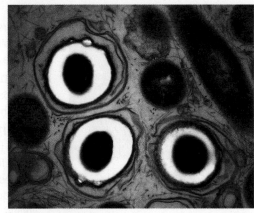

Bacterial spores
David M. Phillips/Science Source

BACTERIAL GROWTH

Bacteria multiply by **binary fission**; that is, they split into two. Each of these two daughter cells then grows to maturity, and each will divide again into two. Bacteria will continue to divide at regular intervals while conditions are suitable for their growth and multiplication. This state is described as the **vegetative state**.

The phase in bacterial growth when bacteria multiply rapidly is called the **logarithmic phase**, or log phase. However, bacteria are not in this phase all the time. If bacteria are not multiplying at all—for example, immediately following removal from a refrigerator—then they are said to be in the **lag phase** If the number of bacteria produced by multiplication equals the number of bacteria dying, then they are in the **stationary phase**. If the number of bacteria decreases and there are more bacteria dying than multiplying—for example, if the food is being cooked at the correct temperature or if there are few or no suitable nutrients—then the bacteria are in their **decline phase**. Depending on the availability of the right nutrients, moisture, and temperature range, bacteria can have longer or shorter phases, or can remain in a phase for a long period of time.

Time/temperature control for safety foods (commonly referred to as TCS foods), are food products that, under the right circumstances, support the growth of harmful bacteria.

Binary fission
CNRI/Science Source

Phases of Bacterial Growth

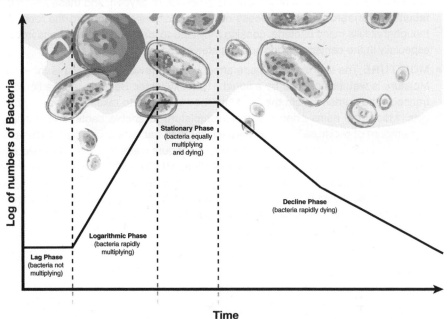

Log of numbers of Bacteria

Stationary Phase
(bacteria equally multiplying and dying)

Logarithmic Phase
(bacteria rapidly multiplying)

Decline Phase
(bacteria rapidly dying)

Lag Phase
(bacteria not multiplying)

Time

Food
Lukas Gojda/Shutterstock

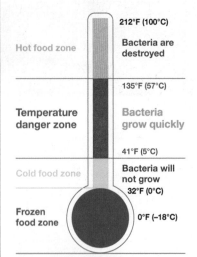

Acidity

Hot food zone	212°F (100°C) **Bacteria are destroyed**
	135°F (57°C)
Temperature danger zone	**Bacteria grow quickly**
	41°F (5°C)
Cold food zone	**Bacteria will not grow** 32°F (0°C)
Frozen food zone	0°F (–18°C)

Temperature danger zone

Moisture

There are six major variables that affect the growth of bacteria. These variables are food, acidity, time, temperature, oxygen, and moisture. The first letter of each variable form the acronym **FAT TOM**.

- **FOOD** To multiply to dangerous levels, bacteria need the right sort of food to support bacterial multiplication. Bacteria get the nutrients they require from food that is high in protein, such as meat, fish, and dairy products.

- **ACIDITY** The amount of acid in food will also affect the growth of bacteria. Food items that are **acidic** don't usually support bacterial multiplication. Acidity has been used for centuries to preserve food. The tartness of some foods, such as citrus, pickles, and sauerkraut, is the result of the food's acidity. Sometimes food acids are added to act as food preservatives and antioxidants. Some common food acids include vinegar, citric acid, malic acid, and lactic acid. **pH** is a unit of measurement that can be used to rate the acidity and alkalinity of food. The pH of a food item is measured on a scale of 1 to 14. Acidic foods have pH values below 7 and **alkaline** foods have pH values above 7; water has a pH value of 7, which is neutral. Most bacteria prefer a pH between 4.6 and 7; that is, mildly acidic foods.

- **TIME** Bacteria also need time at the right temperature in order to multiply to dangerous levels. The time between each bacterial division is known as the **generation time**. The average generation time of the common food-poisoning bacteria, under optimum conditions, is around 20 minutes. However, some food-poisoning bacteria can split into two every 10 minutes. This means that in just a little over an hour and a half, 1,000 bacteria can become over one million.

- **TEMPERATURE** Food-poisoning bacteria generally multiply between 41°F and 135°F (5°C and 57°C)—the **temperature danger zone**. Most food-poisoning bacteria multiply fastest between 70°F and 125°F (21°C and 52°C), especially at body temperature, which is 98.6°F (37°C). Some bacteria, such as *Listeria*, multiply at temperatures below 41°F (5°C). None will multiply at temperatures above 135°F (57°C), but remember that spore-producing bacteria will not be completely destroyed at those higher temperatures. Therefore, if hot food is not kept hot enough or cold food is not kept cold enough, bacteria may begin to multiply again. Cold food should be kept below 41°F (5°C) and hot food should be kept above 135°F (57°C). Bacteria multiply more slowly when food is refrigerated, and freezing food can further slow or even stop the spread of bacteria. However, bacteria in refrigerated or frozen foods become active again when the food is warmed.

- **OXYGEN** Another variable affecting the growth of bacteria is oxygen. Oxygen is normally present in food, except for food subjected to boiling or canning. Some bacteria, called **aerobes**, can multiply only in the presence of oxygen, while others, called **anaerobes**, can multiply only when there is no oxygen. Other bacteria, however, can multiply with or without the presence of oxygen, and these are called **facultative anaerobes**. The process of cooking drives oxygen from the food. Although that kills many bacteria, cooking may lead to ideal conditions for anaerobes, especially in the center of foods such as stews.

- **MOISTURE** The last major variable affecting the growth of bacteria is moisture. Moisture is wetness caused by a liquid—typically water. Bacteria require liquid to transport the nutrients from the food into their cells and to take away waste products. Most food items contain sufficient moisture to enable bacteria to multiply. The amount of moisture in a food item available to bacteria is usually measured in terms of **water activity** (a_w). The water activity of pure water is 1, and most bacteria multiply best in food with a water activity between 0.95 and 0.99, for example, fresh meat. The addition of non-aqueous constituents to water will decrease the a_w to less than 1. As the amount of available water in a food is reduced below this level, the ability of microorganisms to grow is inhibited.

Specifically, TCS foods are foods that are high in protein, have neutral or slightly acidic pH levels, and have high moisture content. TCS foods require strict time and temperature control to prevent the growth of microorganisms. In addition to time and temperature, acidity and water activity play a significant role in bacterial growth in TCS foods.

For example, milk products, meats, and vegetables—which tend to be neutral in pH—will support the growth of microorganisms if not handled correctly. They are considered TCS foods. Yet generally a pH of less than 4.6 will severely restrict or completely stop the growth of harmful bacteria in food. Therefore, foods with higher acidity, such as vinegars, olives, and citrus foods, inhibit the growth of microorganisms and are not designated as TCS foods.

Additionally, foods with high water activity, such as custards, meats, and melons, support the growth of microorganisms and are considered TCS foods. Foods with low water activity, such as breads, dried meats, and jerky, are not.

TYPES OF BACTERIA

Bacteria	Sources	Common Food Vehicles	Symptoms	Onset Time	Duration	Specific Controls
Bacillus cereus	• Starchy foods (emetic toxin) • Meat products (diarrheal toxin)	• Rice • Corn • Potatoes • Meat products	EMETIC: • Nausea • Vomiting DIARRHEAL: • Abdominal cramps • Watery diarrhea	EMETIC: .5 to 6 hours DIARRHEAL: 6 to 15 hours	EMETIC: Less than 24 hours DIARRHEAL: 24 hours	• Cook food according to the proper time and temperature guidelines • Make sure to use proper cooling techniques
Campylobacter jejuni	• Healthy cattle, chickens and birds • Flies • Nonchlorinated water	• Chicken • Raw milk	• Diarrhea • Abdominal pain • Fever • Headache	2 to 5 days	7 to 10 days	• Cook food according to the proper time and temperature guidelines • Prevent cross-contamination of raw meat and ready-to-eat food
Clostridium botulinum	• Soil • Sediment • Intestinal tracts of fish and mammals	• Sausages • Meat products • Canned vegetables • Seafood products	• Weakness, vertigo • Abdominal distention • Double vision • Difficulty speaking and swallowing	18 to 36 hours	Several months	• Pay close attention to time and temperature guidelines • Be careful not to use the contents of damaged cans
Clostridium perfringens	• Soil • Animal and human waste • Dust • Insects	• Stews • Casseroles • Meat pies	• Diarrhea • Abdominal pain	8 to 22 hours	24 hours	• Pay close attention to time and temperature guidelines • Carefully wash raw produce • Ensure food preparation equipment is properly cleaned
Escherichia coli	• Infected workers • Contaminated water	• Undercooked beef • Unpasteurized milk or juice • Contaminated produce	• Severe abdominal cramping • Sudden onset of watery diarrhea, frequently bloody • Sometimes vomiting and a low-grade fever	3 to 5 days	1 to 3 days	• Cook meats and poultry thoroughly • Do not consume raw milk or unpasteurized dairy products • Wash hands after using the bathroom and before preparing or eating food

Bacteria	Sources	Common Food Vehicles	Symptoms	Onset Time	Duration	Specific Controls
Listeria monocytogenes	• Soil • Domestic and feral animals • Raw milk	• Deli meats, hot dogs • Soft cheese • Raw vegetables • Raw meats, poultry and fish	• Flu-like symptoms • Persistent fever	Several days to 3 weeks	Varies	• Cook food according to the proper time and temperature guidelines • Throw away products that have passed their use-by date
Salmonella spp.	• Soil • Sewage • Insects • Feces	• Raw meats • Poultry • Eggs • Milk and dairy products • Fish	• Nausea • Vomiting • Abdominal cramping • Fever	6 to 48 hours	1 to 2 days	• Cook food according to time and temperature guidelines • Prevent cross-contamination of raw meat and ready-to-eat food • Use proper hand washing techniques
Salmonella typhi	• Contaminated food and water • Feces	• Poultry • Raw meats • Raw fruits and vegetables	• High fever • Headache • Lethargy • Abdominal pains • Diarrhea or constipation • Occasionally a rash occursr	1 to 3 weeks, up to 2 months	2 to 4 weeks	• Prevent cross-contamination of raw meat and ready-to-eat food • Use proper hand-washing technique
Shigella spp.	• Infected workers • Contaminated water	• Deli salads • Raw vegetables • Milk and dairy products • Poultry	• Vomiting • Abdominal pain • Diarrhea • Blood in stool	12 to 48 hours	5 to 7 days	• Use proper hand-washing techniques • Carefully wash raw produce
Staphylococcus aureus	• Human skin, nose and hands; boils and cuts • Dust • Raw milk from cows or goats	• Milk and dairy products • Deli salads • Desserts • Custards • Cooked meat and poultry	• Nausea • Vomiting • Abdominal cramping • Diarrhea	30 minutes	2 days	• Use proper hand washing techniques • Cover cuts on arms and hands • Pay attention to time and temperature guidelines
Toxoplasma gondii	• Infected workers, contaminated water	• Raw or undercooked meat and seafood, cat feces	• Asymptomatic in most healthy individuals • Sore lymph nodes and muscle pains, blurred vision, tearing or redness in the eye, sensitivity to light	5 to 23 days	Several weeks	• Cook food according to time and temperature guidelines • Use proper hand-washing techniques • Carefully wash raw produce
Vibrio parahaemolyticus	• Raw and under-cooked fish and shellfish	• Oysters • Clams • Crab • Shrimp • Tuna • Sardines	• Diarrhea • Abdominal cramps • Nausea • Vomiting • Headache • Fever and chills	4 to 96 hours	3 days	• Purchase seafood from reputable suppliers • Cook seafood according to the proper time and temperature guidelines

LESSON 3 | VIRUSES

Learning Objective: Describe the characteristics of viruses.

Viruses are submicroscopic and parasitic, meaning that they are too small to be seen with even a regular microscope, and they need to enter another living cell in order to multiply. Most viruses can survive freezing, and they can contaminate both food and water supplies. Some viruses cause foodborne illnesses; these are classified as infections. Viruses are usually brought into facilities by food handlers who are carriers, or on raw food, such as shellfish, that has been in contact with sewage-polluted water. More often, however, viruses hitchhike on another object, known as a vehicle. The main vehicles are hands; hand-contact surfaces, such as refrigerator handles or faucet taps; clothing and equipment; and food-contact surfaces, such as cutting boards. Viruses are present in the stool or vomit of people who are infected. People who are infected with a virus may contaminate food and drinks, especially if they do not wash their hands thoroughly after using the bathroom. The path along which contaminants are transferred from their sources to food is known as the **route of contamination**.

Common sources of foodborne viruses include:

- Food prepared by a person infected with a virus
- Shellfish from contaminated water
- Produce irrigated with contaminated water

VIRAL REPRODUCTION

Viruses are made up of genetic material that is sealed in a protein coat. Viruses do not reproduce on their own, so they must infect a host cell. Viruses can enter human cells through a watery pathway between proteins known as a protein channel. Once inside the cell, the virus infects the cell with its genetic information. This information causes the infected cell to multiply by making copies of itself. These are finally released when, as a defense mechanism, the original infected cell bursts. These newly produced cells go on to rapidly infect other previously healthy cells.

VIRAL INFECTIONS

The two most common foodborne viruses are:

- Norovirus, which causes inflammation of the stomach and intestines
- Hepatitis A, which causes inflammation of the liver

Norovirus

A **norovirus** infection is commonly called gastroenteritis (also called stomach flu). Each year, it causes around 20 million illnesses and contributes to approximately 65,000 hospitalizations and 700 deaths.[2] Norovirus is the most common cause of foodborne illness outbreaks in the United States.

Norovirus:

- Is found in the feces of infected persons and in contaminated water
- Can be spread by infected food workers who handle food without properly washing their hands after using the restroom
- Causes the symptoms of:
  - Nausea
  - Vomiting
  - Abdominal cramps
  - Diarrhea
- Has an onset time of usually between 24 and 48 hours

- Has symptoms that can last from 24 to 60 hours
- Usually goes away for a complete recovery
- Can be prevented by proper hand-washing techniques
- Spreads easily; infected employees should be prohibited from working in the facility until they have been cleared by a doctor

Hepatitis A

Hepatitis A is a self-limiting disease that does not result in chronic infection. The term *self-limiting* means that the virus "helps" to rid the host of the virus by causing vomiting and diarrhea. Hepatitis A is highly contagious; the process of **transmission**, or spreading of the disease, is usually by the fecal-oral route, through either person-to-person contact or consumption of contaminated food or water. A vaccine is available that will prevent infection from hepatitis A for at least 10 years.

Hepatitis A:

- Is primarily found in the feces of infected persons
- Can be spread from infected food workers to ready-to-eat food, such as:
  - Deli meats
  - Produce
  - Salads
- Can be found in raw shellfish
- Causes the symptoms of:
  - Nausea
  - Vomiting
  - Diarrhea
  - Mild fever
  - Headache
  - General fatigue
  - Flu-like manifestations
- May cause **jaundice**, a yellowing of the skin or eyes, indicating liver malfunction in the infected person
- May exhibit no symptoms at all, especially in children
- Has an onset time of two to six weeks after start of infection
- Has symptoms that may return over the following 36 months
- Can be prevented by proper hand-washing techniques
- Spreads easily; infected employees should be prohibited from working in the facility until they have been cleared by a doctor

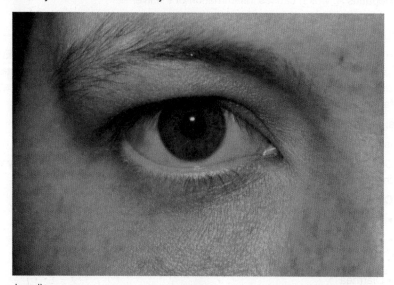

Jaundice
Garry Watson/Science Source

It is important to note that the U.S. Centers for Disease Control and Prevention (CDC) has a published list of infectious and communicable diseases that can be transmitted through food. The list includes diseases caused by both bacteria and viruses. The CDC has found no evidence that **HIV** can be transmitted through food. Because of this, an employee who has tested HIV positive is not a concern unless that person also suffers from a secondary illness as a result of a foodborne infection or intoxication.

PRIONS

Prions are small pathogenic proteins that, like viruses, require a living host in order to multiply. They are best known for causing bovine spongiform encephalopathy (mad cow disease), a fatal infection of the brain in cattle and other livestock. FDA regulations that took effect in 1997 prohibit the use of animal protein in the veterinary feed of cows, sheep, and goats. Prions can't be detected by color, odor, or taste, so the best way to avoid this pathogen is to purchase only meat inspected by the USDA.

Learning Objective: Describe the characteristics of parasites and fungi.

PARASITES

Parasites are plants or animals that live on or in another plant or animal—the host—to survive.

Some attributes of parasites are as follows:

- Range in size from tiny, single-celled organisms to worms visible to the naked eye
- Live and reproduce within the tissues and organs of infected human and animal hosts
- Are often excreted in feces
- May be transmitted through consumption of contaminated food and water
- Can be destroyed by freezing at −4°F (−20°C) or below for 7 days, or −31°F (−35°C) or below for 15 hours

Cryptosporidium parvum and *Giardia intestinalis* are parasites that are spread through water contaminated with the stools of infected people or animals. Foods that come into contact with contaminated water during growth or preparation can become contaminated with these parasites. Food preparers who are infected with these parasites can also contaminate foods if they do not thoroughly wash their hands after using the bathroom and before handling food.

Trichinella spiralis is a type of roundworm parasite that causes **trichinosis**. People may be infected with this parasite by consuming raw or undercooked pork or wild game.

FUNGI

Fungi are biological contaminants that can be found naturally in air, plants, soil, and water. Some fungi are useful to humans. For example, *Saccharomyces cerevisiae*, or baker's yeast, is used to make bread rise and to brew beer. Fungi also aid in the **decomposition** of dead plants and animals. With respect to food safety, the two types of fungi of concern to food managers are **molds** and **yeasts**.

Molds:

- Are threadlike organisms that produce spores
- Are multicellular

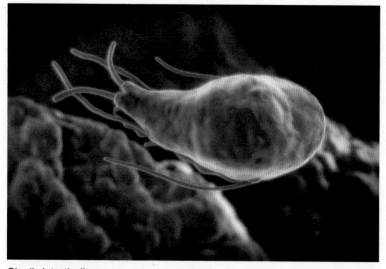

Giardia intestinalis
Giardia intestinalis/Science Source

TYPES OF PARASITES

Bacteria	Sources	Common Food Vehicles	Symptoms	Onset Time	Duration	Specific Controls
Anisakis simplex	• Raw or under-cooked seafood	• Sushi • Squid	• Tingling sensation in throat • Cough up nematode • Abdominal pain	1 hour to 2 weeks	Varies	• Cook seafood properly • Freeze food
Giardia lamblia / Giardia duodenalis	• Unwashed hands • Contaminated water	• Raw vegetables	• Nausea • Abdominal cramping • Vomiting	3 to 25 days	1 to 2 weeks	• Use properly treated water • Follow proper hand washing techniques
Cryptosporidium parvum	• Contaminated soil, food, water or surfaces	• Raw fruits and vegetables	• Watery diarrhea • Coughing • Low-grade fever	2 to 10 days	2 to 4 days	• Use properly treated water • Follow proper hand washing techniques • Thoroughly wash all vegetables and fruits in clean water
Cyclospora cayetanensis	• Contaminated water • Various types of produce	• Raw fruits and vegetables	• Watery diarrhea • Constipation • Nausea • Abdominal cramping • Fever	1 week	Several days to 1 month	• Use properly treated water • Follow proper hand washing techniques • Thoroughly wash all vegetables and fruits in clean water
Trichinella spiralis	• Raw or under-cooked food	• Wild game • Pork	• Nausea • Vomiting • Diarrhea • Fever • Abdominal pain • Headache • Eye swelling • Aching joints • Muscle weakness	1 to 2 days	2 to 8 weeks	• Cook all pork and wild game properly

- Can be seen with the naked eye
- Can cause allergic reactions and respiratory problems
- Can spoil food
- Are not destroyed by freezing, although freezing can prevent or reduce the growth of mold
- Can survive cooking
- Grow well in food with high acidity and low water activity
- Grow at refrigerated temperatures below 41°F (5°C)
- Grow in bread stored under moist conditions or bread that sits for too long
- Can be distinguished on the surface of food items by their fuzzy appearance, green or black in color

Mold is mostly a **spoilage organism**, which means it damages the nutrition, texture, and flavor of the food, making it unsuitable to eat. Some molds cause allergic reactions and respiratory problems. In addition, a few molds, in the right conditions, produce mycotoxins, poisonous substances that can cause serious illness. One type of **mycotoxin** is aflatoxin. **Aflatoxins** that result from mold growth can develop in peanuts or other crops that are stored in a moist environment. Moldy peanuts or any other crops should be disposed of and not eaten; serious illness can result. Cooking cannot kill aflatoxins.

Some mold is a natural part of the food item, as is the case with certain types of cheeses. Where molds are not natural, as in hard cheeses, the Food and Drug Administration (FDA) recommends cutting away moldy areas one inch from where they occur.

Yeasts:

- Spoil food at a very quick pace
- Require oxygen to grow
- Produce carbon dioxide and alcohol
- Can result in a distinct taste or smell of alcohol
- Can appear pink in color
- May be slimy
- May bubble
- Are sensitive to normal cooking temperatures and are destroyed by cooking

Yeasts are used in the manufacture of foods such as bread, beer, and vinegar; however, they do cause spoilage of foods such as jam, fruit juice, honey, meats, and wines. If any of these foods appear to be spoiled by yeast, they should be discarded. Yeasts are very rarely implicated in causing illness, but they will often lead to customer complaints.

LESSON 5 | CHEMICAL CONTAMINATION

Learning Objective: Identify best practices for preventing chemical contamination.

Chemical contamination is the presence of unwanted chemical components in food or the food environment.

CHEMICALS

Chemicals such as pesticides, cleaning agents, and refrigerants are commonly used in food preparation areas and are a common source of chemical contamination.

- When using any chemical, follow all EPA-registered label use instructions. This includes strictly following any warnings as well as directions for proper chemical mixing.
- Never use a stronger chemical than is needed to achieve the desired result.
- Food should also be stored, covered, or otherwise protected when using chemicals. Because of the difficulty of protecting food during operating hours, a good rule is to try to use toxic chemicals, such as pesticides, after operating hours when all exposed food is properly stored.
- Consider using a licensed professional to apply chemicals such as pesticides.

To prevent chemical contamination through improper storage, use the following guidelines:

- Chemicals should have a separate storage area—never store them with food, utensils, or other food preparation equipment.
- Store chemicals in their original EPA-registered label use containers along with instructions for use. If chemicals must be transferred to another container, ensure that container is properly labeled and stored.
- Any containers or utensils used with chemicals should never be used for food storage or preparation.

Pesticide residue
Crown Copyright courtesy of Central Science Laboratory/Science Source

TOXIC METAL POISONING

Toxic metal poisoning can occur when acidic foods such as tomatoes are stored or cooked in aluminum or copper containers. The acid in these foods leaches the toxic metals from the container into the food.

Examples of foods with high acid levels that can create this problem are:

- Grapefruit
- Mayonnaise
- Vinegar
- Oranges
- Pineapple

Toxic metal poisoning causes both unpleasant-tasting food and illness. To prevent this type of poisoning, never store food in galvanized metal. Take steps to ensure that all storage and preparation equipment, including utensils, are food grade.

Toxic metal poisoning can also occur through improper installation or maintenance of carbonated-beverage dispenser systems.

This type of toxic metal poisoning occurs if carbonated water flows back through copper supply lines. As occurs with acidic foods, the carbonated water will leach copper from the lines and contaminate the beverage. To prevent carbonated beverages from becoming contaminated, ensure that all beverage systems are professionally installed and maintained and are fitted with a backflow prevention device, which is discussed in more detail in the plumbing section of chapter six.

PREVENTING CHEMICAL CONTAMINATION

To prevent chemical contamination that may come through the food supply, ensure that:

- Suppliers are following all appropriate regulations for raising, slaughtering, and harvesting their product. This may include random quality checks and appropriate penalties.
- Any lubricants or oils used with kitchen equipment are food grade.
- Chemicals used for cleaning and sanitizing are approved for use in a food facility.
- Employees are properly trained in cleaning and sanitizing practices. This includes properly washing all appropriate foods as well as thoroughly washing their own hands after using any chemicals.

Improperly storing foods can also lead to chemical contamination.

- Never store food in a container that was previously used to store chemicals.
- It is always best to keep chemicals in their original containers, but if they come in bulk and need to be placed into a separate container for ease of use, the containers must be labeled with the type of chemical.
- Store chemicals in a separate room or area away from food. In addition to cleaning and sanitizing chemicals, this includes first-aid supplies and personal employee hygiene items.

LESSON 6 | NATURAL TOXINS

Learning Objective: Identify the natural toxins that can cause foodborne illness.

NATURAL TOXINS

Another type of chemical contaminant is the natural toxin. Natural toxins that can cause food-borne illness come from:

- Fish or shellfish, which may feed on algae that produce toxins, leading to high concentrations of toxins in their bodies.
- Some types of fish, including tuna and mahi mahi, that may be contaminated with bacteria that produce toxins if the fish are not properly refrigerated before they are cooked or served.
- Certain types of wild mushrooms.
- Red kidney beans or fava beans when consumed raw or undercooked.
- Certain plants, including:
  - Deadly nightshade
  - Death cap mushrooms
  - Daffodil bulbs
  - Rhubarb leaves

Common symptoms of illness caused by toxic plants include:

- Nausea
- Vomiting
- Abdominal pain

The adverse symptoms of plant toxins generally have an onset time of one to six hours. Fish and shellfish toxins have very different symptoms.

Deadly nightshade
Scott Camazine/Science Source

CIGUATOXIN

The gonads, liver, and intestines of some fish are highly toxic and may result in food poisoning. Poisoning by **ciguatoxin** is caused by eating certain fish which have toxins and cause sea-food poisoning. In the southern Florida, Bahamian, and Caribbean regions, snapper, grouper, and mackerel are examples of fish that produce toxins.

Symptoms of food poisoning by natural fish toxins include:

- Tingling of the fingers
- Disturbance of vision
- Paralysis
- Nausea
- Vomiting
- Diarrhea

To prevent illnesses from occurring, only purchase fish from reputable suppliers. When the fish is delivered, check that its temperature is 41°F (5°C) or lower. Never accept deliveries that show signs of refreezing, such as ice crystals or freezer burn.

SCOMBROTOXIC FISH POISONING

Scombrotoxic fish poisoning is caused by **scombrotoxins** that accumulate in the body of certain fish (for example, tuna, mahi mahi, bluefish, sardines, mackerel, amberjack, and anchovies) during storage. It is likely to occur when temperatures rise above 39.2°F (4°C).

Symptoms include:

- Headache
- Nausea
- Vomiting
- Abdominal pain
- A rash on the face and neck
- A burning or peppery sensation in the mouth
- Sweating
- Diarrhea

These symptoms may last up to eight hours. Again, purchase fish from only reputable suppliers. Also check for signs of time/temperature abuse.

SHELLFISH TOXINS

Shellfish food poisoning may result from the consumption of mussels and other bivalves that have fed on poisonous algae or plankton. These toxins have no odor and no taste, and cannot be destroyed by freezing or cooking.

Symptoms caused by shellfish poisoning include:

- Headache
- A floating feeling
- Dizziness
- Lack of coordination
- Tingling of the mouth, arms, or legs

The best way to protect the establishment and still offer shellfish to the customer is to purchase shellfish only from reputable suppliers. Seafood certification tags that list where and when shellfish were harvested are required on all shellfish, and must be kept on file on the premises for at least 90 days.

Contaminant	Common Food Vehicles	Specific Controls
Scombrotoxin	Primarily associated with tuna fish, mahi-mahi, blue fish, anchovies, bonito, mackerel; also found in cheese	Check temperatures at receiving; store at proper cold holding temperatures; Buyer specifications: obtain verification from supplier that product has not been temperature abused prior to arrival in facility.
Ciguatoxin	Reef fin fish from extreme SE US, Hawaii, and tropical areas; barracuda, jacks, king mackerel, large groupers and snappers	Purchase fish from approved sources. Fish should not be harvested from an area that is subject to an adverse advisory.
Tetrodoxin	Puffer fish (Fugu; Blowfish)	Do not consume these fish.
Mycotoxins *Aflatoxin*	Corn and corn products, peanuts and peanut products, cottonseed, milk and tree nuts such as Brazil nuts, pecans, pistachio nuts and walnuts. Other grains and nuts are susceptible but less prone to contamination.	Check condition at receiving; do not use moldy or decomposed food.
Patulin	Apple juice products	Buyer specification: obtain verification from supplier or avoid the use of rotten apples in juice manufacturing.
Toxic mushroom species	Numerous varieties of wild mushrooms	Do not eat unknown varieties or mushrooms from unapproved source.
Shellfish toxins *Paralytic shellfish poisoning (PSP)*	Molluscan shellfish from NE and NW coastal regions; mackerel, viscera of lobsters and Dungeness, tanner and red rock crabs	Ensure molluscan shellfish are from an approved source and properly tagged and labeled.
Diarrhetic shellfish poisoning (DSP)	Molluscan shellfish in Japan, western Europe, Chile, NZ, eastern Canada	
Neurotoxin shellfish poisoning (NSP)	Molluscan shellfish from Gulf of Mexico	
Amnesic shellfish poisoning (ASP)	Molluscan shellfish from NE and NW coasts of NA; viscera of Dungeness, tanner, red rock crabs and anchovies.	

*Although the 2009 FDA Food Code catagorizes toxins as chemical contaminants, in food safety management, they are traditionally catagorized as biological hazards, so that is how we have decided to present them here.

Learning Objective: List the major food allergens.

Allergens are a third type of chemical contaminant. An **allergen** is any substance that can cause an substance reaction in some people, which happens when the immune system sees the substance as foreign or dangerous. Nearly seven million Americans suffer from food allergies. Food allergies affect two percent of adults and five percent of infants and children in the United States.[3] Food allergies can be deadly; proper labeling is an important element of food safety.

The Food Allergen Labeling and Consumer Protection Act (FALCPA) took effect in 2006. This law requires all food containing a major food allergen, or a food ingredient that contains protein derived from a major food allergen, to be clearly labeled.

FOOD ALLERGY SYMPTOMS

A **food allergy** is the manifestation of the body's immune system responding to a protein that it mistakenly believes is harmful. The body releases **histamines** as an immune response to protect itself against the allergen. Depending on the person, the immune response could be immediate and severe, or it could be delayed for a period of time, with milder symptoms.

Symptoms of food allergies include:

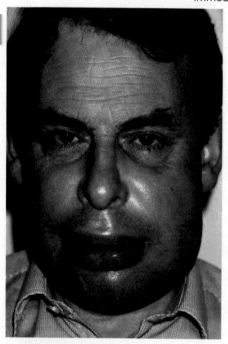

Food allergy swelling
Dr. P. Marazzi/Science Source

- A tingling sensation in the mouth or throat
- Itching in and around the mouth, face, and/or scalp
- Swelling, including swelling of the tongue, throat, face, eyes, hands, and feet
- Difficulty breathing, including wheezing or shortness of breath
- Rash or hives
- Nausea and/or vomiting
- Abdominal cramps
- Diarrhea
- Loss of consciousness

A serious allergic reaction that is rapid in onset and life-threatening is called an **anaphylactic reaction**. Food allergies are believed to be the leading cause of anaphylaxis outside of a hospital setting. Because food allergies can cause death, it is important to train staff to recognize an allergic reaction so that proper medical treatment can be administered.

Food allergies cannot be cured. The only way to prevent an allergic reaction is to avoid the food that triggers it.

COMMON FOOD ALLERGENS

Eight major foods, known as **"the Big Eight,"** account for 90 percent of all food allergies in the United States:

- Milk
- Eggs
- Fish
- Shellfish
- Tree nuts
- Peanuts
- Wheat
- Soy

Food allergens
Evan Lorne/Shutterstock

These major allergens must be properly labeled under the 2006 Food Allergen Labeling and Consumer Protection Act (FALCPA).

FALCPA applies only to packaged, FDA-regulated foods. The labeling requirement does not cover food service and retail establishments. Food placed in a wrapper or container in response to a customer order (such as a box for a sandwich) does not have to be labeled.

In addition to naturally occurring food allergens, food additives can also bring about life-threatening allergic reactions. Ensure that food additives are used in strict compliance with the manufacturer's instructions and that they are properly labeled. MSG, or monosodium glutamate, is a food additive that can cause allergic reactions in some people. Symptoms tend to occur within one hour of eating three grams or more on an empty stomach. Severe, poorly controlled asthma may predispose someone to a reaction to MSG.

PREVENTING ALLERGIC REACTIONS

As with many other elements of food safety, implementing appropriate policies and training can help to prevent allergic reactions.

- Train staff to know which menu items contain major food allergens and to recognize the symptoms of a reaction.
- Ensure that at least one person on each shift is thoroughly familiar with all the food ingredients used.
- Train cooking staff to be able to prepare menu items that are allergen-free if a customer requests it.
- Impress upon staff that people suffering from food allergies must totally avoid the food in question, as even tiny amounts can cause a severe reaction.
- Use good food preparation practices.
- Avoid cross-contact with an allergenic food.

Other good practices for preventing allergic reactions include:

- Carefully reading all labels
- Labeling allergenic items on the menu
- Using a separate prep area for items containing allergenic ingredients
- Carefully cleaning tables and utensils that have come into contact with an allergen
- Serving sauces on the side
- Avoiding product substitutions for menu items
- Being alert if a customer indicates they have an allergy
- Having an emergency procedure in place to handle allergic reactions and calling for treatment when necessary

Acidic   Having a pH level less than 7, as do foods such as vinegar and tomatoes.

Aerobe   An organism that requires oxygen to multiply. Also referred to as an aerobic organism.

Aflatoxins   A type of mycotoxin found in moldy peanuts, seeds, and spices that cannot be killed by cooking.

Alkaline   Having the opposite chemical property from acids. Alkaline products have a pH level greater than 7.

Allergen   Any substance that can cause an allergic reaction in some people, when their immune system sees the substance as foreign or dangerous.

Anaerobe   An organism that requires the absence of oxygen to multiply. Also referred to as an anaerobic organism.

Anaphylactic reaction   A severe allergic reaction affecting the whole body, often within minutes of eating the food, which may result in death. Also referred to as anaphylaxis.

Bacillus cereus (B. cereus)   Intoxication-causing bacteria commonly found in starchy foods and meat products. This type of bacteria produces two types of toxins: emetic and diarrheal.

Bacteria   Single-celled microorganisms with rigid cell walls that multiply by dividing in two; that is, by binary fission. Some bacteria cause illness, and others cause food spoilage.

Bacterial spores   A resistant resting phase of bacteria, protecting them against adverse conditions such as high temperatures.

Big Eight allergens   The major food allergens: milk, eggs, fish, shellfish, tree nuts, peanuts, wheat, and soy. These foods account for about 90 percent of all food allergies in the United States.

Binary fission   The method by which bacteria multiply; they split into two.

Biological contamination   Food contamination by microorganisms, including bacteria, viruses, parasites, and fungi.

Campylobacter jejuni   Infection-causing bacteria found on raw poultry and in contaminated water.

Carrier   A person who harbors, and may transmit, pathogenic organisms with or without showing any signs of illness.

Chemical contamination   The contamination of food by chemical substances such as pesticides and cleaning solutions. Includes contamination by natural toxins and allergens.

Ciguatoxin   A toxin found in some tropical coral reef fish. The toxin causes the following symptoms when consumed: nausea, vomiting, diarrhea, muscular weakness, numbness in extremities, and possibly respiratory arrest.

Clostridium botulinum   An intoxication-causing bacteria commonly found in soil and therefore in products that come from soil such as root vegetables. It is anaerobic, which means it grows without oxygen. Because there's no need for oxygen, C. botulinum can also be found in improperly canned food.

Clostridium perfringens   A bacteria that causes mild infection from toxin-producing spores. It is anaerobic and can be found in soil, animal and human waste, dust, insects, and raw meat.

Contamination   The presence of physical, chemical, or biological matter in or on food or in the food environment.

Cross-contamination   Cross-contamination occurs when bacteria from contaminated foods (usually raw) transfers to other foods by direct contract, drip, or indirect contact.

Cryptosporidium parvum   A parasite found in soil, food, or water, or on surfaces that have been contaminated with infected human or animal feces.

Decline phase   The period in the life cycle of bacteria during which more bacteria are dying than multiplying, leading to an overall decrease in their number.

Decomposition   The process of decay, or breaking down of organic matter.

Emetic   Causing vomiting.

Endotoxin   A toxin present in the cell wall of many bacteria that is released upon the death of the bacteria.

Escheria coli (E. coli)   A bacteria found in the intestines of mammals. It can be found in ground beef and contaminated produce.

Exotoxin   A toxin produced during the multiplication of some bacteria. They are highly toxic proteins and are often produced in food.

Facultative anaerobe   An organism that can multiply with or without the presence of oxygen.

FAT TOM   The acronym that represents the conditions that support the rapid growth of bacteria. These conditions are food, acidity, time, temperature, oxygen, and moisture.

Food allergy   An identifiable immunological response to food or food additives, which may involve the respiratory system, the gastro-intestinal tract, the skin, or the central nervous system.

Food spoilage bacteria   Bacteria that diminish food quality, but rarely cause serious illness.

Fungi   Biological contaminants that can be found naturally in air, plants, soil, and water. Fungi can be small, single-celled organisms or larger, multicellular organisms, and include molds and yeasts.

Generation time   The time between bacterial divisions.

Germinate   Also known as germination; the development or growth of microorganisms.

Giardia intestinalis   Also, Giardia duodenalis, Giardia lamblia; a parasite found in contaminated water, raw fruits, and vegetables.

Hepatitis A   A virus primarily found in the feces of infected persons. It is spread from infected food workers to ready-to-eat food, including deli meats. It can also be spread to produce and salads and can be found in raw shellfish.

Histamine   A naturally occurring substance produced in the body as an immune response to an allergen.

HIV   A retrovirus spread through blood and bodily fluids. The CDC has found no evidence that HIV can be transmitted through food.

Infection   A disease caused by the release of endotoxins in the intestine of the affected person.

Infectious   Communicable; tending to spread between people.

Intoxication   An illness caused when bacteria produce exotoxins that are released into food.

Jaundice   A yellowish discoloration of the skin and eyes, indicating liver malfunction and illness.

Lag phase   The period in the life cycle of bacteria during which bacteria are not multiplying at all.

Listeria monocytogenes   Infection-causing bacteria naturally found in soil, raw vegetables, and milk that has not been properly pasteurized. It is associated with certain ready-to-eat foods, such as deli meats and hot dogs.

Logarithmic phase   Also called log phase; the period during bacterial growth in which bacteria multiply rapidly.

Microorganisms   Organisms, or living things, such as bacteria, viruses, fungi, and parasites that are too small to be seen with the naked eye. These microorganisms may contaminate food and cause foodborne illness.

Mold   Microscopic fungi that produce threadlike filaments; mold can be black, white, or of various colors.

Mycotoxins   Poisonous substances produced by certain fungi.

Norovirus   Found in the feces of infected persons. Can also be found in contaminated water. Norovirus is the most common cause of viral gastroenteritis in humans.

Onset time   The period between eating contaminated food and the first signs of illness.

Parasite   An organism that lives and feeds in or on another living creature, known as a host, in a way that benefits the parasite and disadvantages the host.

Pathogen   Disease-producing organism.

pH   An index used as a measure of acidity/alkalinity, measured on a scale of 1 to 14. Acidic foods have pH values below 7 and alkaline foods have values above 7; a pH value of 7 is neutral.

Physical contamination   When any foreign object is in or on a food and presents a hazard or nuisance to those consuming it.

Prions   Small pathogenic proteins that, like viruses, require a living host to multiply.

Ready-to-eat (RTE) food   Food that is going directly to the consumer without further cooking or preparation to kill potential dangerous microorganisms.

Route of contamination   The path along which contaminants are transferred from their sources to food.

Salmonella, spp.   Several species of infection-causing bacteria commonly found in raw poultry, eggs, raw meat and dairy products. It has also been found in ready-to-eat food that has come into contact with infected animals or their waste.

Scombrotoxin   A toxin that forms when certain fish aren't properly refrigerated before being processed or cooked. Examples of fish that can form the toxin if they start to spoil include tuna, mahi mahi, bluefish, sardines, mackerel, amberjack, and anchovies.

Shigella, spp.   Several species of bacteria found in the feces of people with Shigellosis. It can be found in ready-to-eat foods such as greens, milk products and vegetables and also in contaminated water. The most common method of transmission is cross contamination. Flies can also be carriers of this type of bacteria.

Spoilage organism   An organism that damages the nutrition, texture, and flavor of the food, making it unsuitable to eat.

Spore   A resistant resting phase of bacteria, protecting them against adverse conditions such as high temperatures.

Staphylococcus aureus   An intoxication-causing bacteria commonly found on the skin, nose and hands of one out of two people. It is transferred easily from humans to food when people carrying the bacteria handle the food without washing their hands. This bacteria also produces toxins that multiply rapidly in room-temperature food.

Stationary phase   The period in the life cycle of bacteria during which the number of bacteria produced is equal to the number of bacteria dying.

Temperature danger zone   The temperature range at which most foodborne microorganisms rapidly grow. The temperature danger zone is 41°F to 135°F (5°C to 57°C).

Time/temperature control for safety foods (TCS)   Products that under the right circumstances support the growth of microorganisms that cause foodborne illness.

Toxic   Directly poisonous; affected by a toxin or poison.

Toxic metal poisoning   The leaching of certain poisonous metals, such as aluminum or copper, into acidic foods being prepared with pots and/or utensils of those metals.

Toxoplasma gondii   Intracellular parasites. According to the CDC, this parasite is the second leading cause of death from foodborne illness in the U.S. The illness it causes, toxoplasmosis, can be serious or deadly, particularly for babies infected in the womb and people with weak immune systems.

Transmission   The process of spreading (as in a disease or infection) from person to person.

Trichinella spiralis   An intestinal roundworm that is found in wild game animals and in undercooked pork. The larvae of the *Trichinella spiralis* can move throughout the body, infecting various muscles and causing the infection trichinosis.

Trichinosis   An infection caused by *Trichinella spiralis*.

Vegetative state   The condition in which bacteria divide at regular intervals due to surroundings suitable for their growth and multiplication.

Vibrio parahaemolyticus   An infection causing bacteria commonly associated with raw or partially cooked oysters.

Viral gastroenteritis   The swelling or inflammation of the stomach and intestines from a virus, leading to diarrhea and vomiting.

Viruses   Submicroscopic pathogens that multiply in the living cells of their host.

Water activity   A measure of the moisture in food available to microorganisms. It is represented by the symbol a_w. Most bacteria multiply best in food with a water activity of between 0.95 and 0.99.

Yeasts   Single-celled microscopic fungi that reproduce by budding and that grow rapidly on certain foodstuffs, especially those containing sugar.

REFERENCES

[1] Norovirus overview. (2013, July 26). Retrieved from http://www.cdc.gov/norovirus/about/overview.html

[2] National advisory committee on microbiological criteria for foods. (2014, November 17). Retrieved from http://www.fsis.usda.gov/wps/wcm/connect/b53b2da4-574b-4c70-a354-ac6f0ad20dcf/NACMCF-Norovirus-Report-111714.pdf?MOD=AJPERES

[3] Food allergy facts and statistics for the U.S. (2015). Retrieved from https://www.foodallergy.org/document.doc?id=194

ASSESSMENT QUESTIONS

1. The presence of unwanted physical, chemical, or biological matter in or on food or in the food environment is called:
   a. Foodborne illness
   b. Contamination
   c. Infection
   d. Pathogen

2. An infection is associated with:
   a. Exotoxins
   b. Endotoxins
   c. Spores
   d. Intoxications

3. Bacteria multiply by:
   a. Logs
   b. Lag phases
   c. Generation
   d. Binary fission

4. The risk of physical contamination can be reduced by:
   a. Using shields to cover exposed lightbulbs
   b. Applying pesticides after hours when the food is put away
   c. Storing acidic foods in copper containers
   d. Storing chemicals away from food

5. What does *Clostridium botulinum* form to protect itself from refrigerated temperatures?
   a. Prions
   b. Spores
   c. Logs
   d. Fission

6. Illness from norovirus is commonly called:
   a. Hepatitis
   b. Botulism
   c. Gastroenteritis
   d. Shellfish poisoning

7. Salmon can form toxins when it is:
   a. Dripped on by red meat
   b. Touched by a food handler with poor hygiene
   c. Farm raised
   d. Left out of the refrigerator for too long

8. *Cryptosporidium parvum* is a:
   a. Bacterium
   b. Parasite
   c. Virus
   d. Mold

9. Bacterial growth is affected by the amount of _____ in a food.
   a. Vitamin A
   b. Minerals
   c. Protein
   d. Chemicals

10. Which of the following is one of the Big 8 allergens?
    a. Soy
    b. Shellfish
    c. Chocolate
    d. Tree nuts

QUESTIONS FOR DISCUSSION

1. List some ways that food can be contaminated prior to receiving it in the retail facility.

2. What do norovirus and hepatitis A have in common? How do their symptoms differ?

3. Discuss some reasons why parasites are a greater problem in Third World countries.

4. What are some foods with high acid levels that can cause chemical contamination?

5. Make a list of menu items that contain "hidden" allergens; that is, those that are not directly obvious or in menu descriptions.

6. Describe foods that can sometimes contain natural toxins.

7. Pick three types of bacteria and explain how they can cause foodborne illnesses.

CHAPTER FOUR

PEST CONTROL

Food pests can have a serious effect on the reputation and profitability of food service facilities, and on the health of workers and customers. Pests can enter food facilities and spread disease, and can cause illness or even death. Fortunately, pest management programs that eliminate and control pests within a facility can easily be put into practice. It is the food manager's responsibility to ensure that a facility-appropriate pest management program is developed and implemented. The goal of a successful pest management plan is to prevent pests from entering a facility and, if found, to safely eliminate them.

After reading this chapter, you should be able to:

- Identify the common pests that can compromise food safety.
- Describe the three basic goals of integrated pest management.

- Explain the best practices for preventing pests within a food establishment.
- Describe the proper use and storage of pesticides.

46

Cockroach
jonesmarc/iStock photo

LESSON 1 | PESTS

Learning Objective: Identify common pests that can compromise food safety.

Common **pests** found in the food industry include:

- Insects (for example, flies, wasps, moths, cockroaches, book lice, silverfish, and ants)
- Rodents (for example, rats and mice)
- Birds (for example, pigeons and sparrows, especially in outside eating areas)
- The occasional small animal, such as a stray cat, around the dumpster

Food managers should focus pest control efforts on preventing the spread of disease. Cockroaches, rodents, and flies are of special concern because they can cause illness. It is important to know how to identify these pests and where to look for them.

COCKROACHES

There are many different species of cockroaches, ranging in size from ½ inch to 1¾ inches long. Cockroaches are golden/dark brown to black in color. Their bodies are flat and broad, with long antennae. They dislike light, prefer to be out at night, and will scatter if disturbed. They have a strong, oily odor. Their feces look like specks of black pepper.

Cockroaches prefer undisturbed, warm, moist places. Egg casings are an indicator of their presence and important to look for, as one egg can have 30 or more young. Cockroaches are found around sinks, drains, moist boxes or bags, and wet flooring. Other places to look for cockroaches include molding strips, furniture, inside equipment, and wood-burning stoves. Finding cockroach feces or crushed body parts is a sure sign of their presence.

Cockroaches have been identified as carriers of *Salmonella* and the poliomyelitis virus. They are responsible for food poisoning and diarrhea, which can be bloody. Cockroach residue is allergenic, and it can cause asthma symptoms in those sensitive to their residue.

RODENTS

Mice and rats are dangerous rodents to have on-site, and food workers must make sure that rodents are not in the workplace. Rodents can cause serious illness, including Lyme disease, salmonellosis, leptospirosis, and hantavirus pulmonary syndrome. The symptoms of these illnesses include arthritis, fever, diarrhea, flu, respiratory problems, and kidney failure.

Rats and mice are relatives of the beaver. They are gnawing animals that chew entry through most materials. Good indicators of mice and rats in the workplace are their feces, which are brown, rice-shaped pellets. Small, round entry holes or tracks across dusty surfaces near food and water supplies may be an indicator that rodents are present. Mice can squeeze through a hole as small as ¼ inch. Shredded paper, cardboard, cloth, or fur may be signs of rodent nests.

Rat
craftvision/iStock photo

Rats and mice have many of the same physical features, but differ in size. Mice are small mammals with hairless tails and large ears, while rats can be up to nine inches in length. Mice and rats range in color from light to dark brown. Both rats and mice are good climbers, jumpers, and swimmers, and they are typically nocturnal. Under black light, rodent pathways will appear as fluorescent trails due to their urine.

Ants

Robert and Jean Pollock/Science Source

Fly

Pascal Goetgheluck/Science Source

FLIES AND ANTS

Flies are a notorious health problem, and their bodies are covered with debris. If a fly has been near animal waste before entering a facility, that waste will be brought inside. Flies can only ingest liquids, so they vomit onto food prior to eating it in order to dissolve the food. Food managers must keep doors and windows closed or protected, either with screens or by installing air curtains, to prevent flies from entry.

Ants enjoy sweets, so workers should regularly clean up areas where juice, sugar, and fruit are found. Ants clean their antennae and feet frequently, so they do not normally carry disease. However, the premises should be kept free of ants to maintain a healthy and clean environment.

ANIMALS

Stray animals, wildlife, and birds can be a nuisance in outdoor dining, or if they enter a facility. These animals can also carry disease and plague. The best way to prevent animals in outdoor areas is maintenance and cleanliness of the surroundings.

Service animals accompanied by a disabled employee or person may enter food establishments in customer areas. Service animals are not allowed in areas generally designated for employees or food preparation. As long as a health or safety hazard will not result from the presence or activities of the service animal, they are welcome in the food facility.

LESSON 2 | INTEGRATED PEST MANAGEMENT

Learning Objective: Describe the three basic goals of integrated pest management.

Integrated pest management, or IPM, approaches pest control with a wide range of practices to prevent or solve pest problems.

Pest management is essential in order to avoid:

- Contamination and food waste
- Foodborne illnesses and foodborne illness outbreaks
- Damage to equipment and premises—for example, fires caused by gnawed electric cables
- Loss of customers and profits caused by selling contaminated food
- Loss of staff who refuse to work in infested premises or facilities
- Fines and closure of the establishment

Successful implementation of IPM balances prevention with control. Prevention works to keep pests out, and control removes those that get in. IPM recognizes that reliance solely on pesticides to eliminate pests after they enter an establishment can permit an infestation to get out of hand before it is noticed. The strength of IPM is in preventing an occurrence in the first place.

Creating goals and communicating them to staff promotes successful implementation of pest prevention. Goals help educate employees about their responsibilities under the facility's IPM, and they convey the actions needed to prevent infestations. All IPM plans have three basic goals:

1. Prevent pest entrance to the facility.
2. Prevent breeding by removing any food or water sources that could be shelter areas to pests.
3. Work with a pest control operator (PCO) to control any infestations that occur.

Poor pest management can lead to fines and closure of the establishment.
Radius/SuperStock

Use a professional pest control operator.
RubberBall/SuperStock

A **pest control operator (PCO)** is a person licensed to control pests. A PCO must take and pass an examination administered by the state to receive a license. In order to maintain this license, operators must attend training on a continual basis.

An IPM team is made up of food service employees, primarily the establishment manager, kitchen staff, and servers. The facility's PCO is also a valued team member. The PCO's role is to assist and support the facility team at each step in the IPM process. The PCO supports the team by helping the establishment team members:

- **Look for signs of pests.** The PCO has spent more time with pests than other team members. This knowledge is helpful in learning about which pests might be in a facility and finding out how they got there.

- **Identify pests.** If found, the PCO can help the team determine which pests are a serious problem, which are a minor problem, and what action to take and when.

- **Determine a control method.** Control methods are one of the PCO's greatest contributions to the team. Would a physical method, such as traps, barriers, and sticky pads, or a pesticide best control a pest problem? What would improve the cleaning schedule?

- **Evaluate progress.** If pests are resistant to a specific control that is recommended, is there a better control? If not, what are other alternatives to avoid future occurrences?

- **Learn pest control practices.** Once knowledgeable about a facility, a PCO can help team members decide when to handle a situation themselves and when to call the PCO.

As a result of their initial training, on-the-job experience, and continuing education, PCOs have the skills necessary to assist food managers in developing successful IPM programs. PCOs can help food managers:

- **Quickly identify issues.** PCOs know when a pest problem exists, how it will progress, and the most efficient means to stop it. Once a problem is controlled, they can recommend countermeasures to prevent future infestations.

- **More efficiently select treatments.** PCOs are up to date on pest control products, equipment, and techniques. PCOs know what kind of chemicals to use and how much is needed. They know that pesticides are not always the best option.

- **Safely apply treatments.** Only a PCO should apply pesticides in the form of sprays, gases, dusts, and fumigants. The PCO will ensure that no food preparation is in progress and that all food is secure before applying pesticides.

- **Save money.** The right treatments, at the right time and in the right amount, will prevent overuse of pesticides as well as retreatments. Substantial cost savings can be realized by minimizing the use of pesticides in a food service facility.

- **Respond quickly.** The PCO can address and correct an infestation so that food service staff can focus on preparing and serving food.

- **Provide advice and recordkeeping.** As a member of a facility's pest control team, the PCO can provide good housekeeping tips, conduct inspections, and maintain records of the facility's overall pest control effort.

Save money with the proper treatments.
Nicholas Eveleigh/Exactostock-1491/SuperStock

Selecting a PCO from an ad or a referral is a good way to get started, but it shouldn't be the only approach. When selecting a PCO, food managers should use all resources at their disposal. Some ways food managers can select a quality PCO include:

- **Talking to other food service managers in the area.** Find out which pest management company or individuals they use and how satisfied they are with their service.

- **Dealing only with qualified and licensed pest management companies or individuals.** Ask which license or licenses they hold, ask to see them, and make sure that they are up to date. Inquire into a candidate's membership of professional organizations. Check online for membership in the National Pest Management Association.

- **Finding out if the pest management company has liability insurance.** Damages to facility and furnishings can happen during pest management. Ask for verification of liability coverage.

- **Asking if their work is guaranteed.** If so, what is covered and for how long? Thoroughly review any service contracts offered and make sure they meet all needs.

- **Taking time to make a decision.** Seek value, not just cost. Does the person meet all of the facility's needs? Is the person trustworthy?

Learning Objective: Explain the best practices for preventing pests within a food establishment.

CLEANING SCHEDULE

Maintaining a regular cleaning schedule deters pests from entering the facility. Pests are attracted to places where food and water are readily available. Additionally, they find dimly lit, moist, warm locations good places to live and breed. Pests will not thrive in dry, well-lit areas where food is scarce.

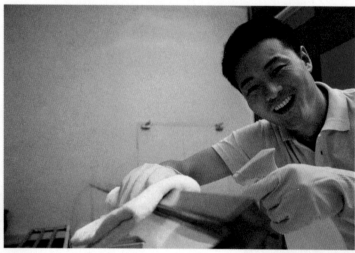

Maintain a regular cleaning schedule.
Blend Images/SuperStock

To prevent pests, the facility's cleaning schedule should include:

- A complete list of items to clean: floors, drains, walls, ceilings, HVAC, doors, windows, equipment, work surfaces, toilets, rest areas, dining areas, and so on
- Precautions to take for each area or item that could be at risk
- Inspections to monitor the effectiveness of the schedule: Are scheduled times working? Are more frequent inspections necessary? Are there other things that need inspection?

WASTE MANAGEMENT

Good waste management includes proper refuse and waste disposal. Be sure to focus on the following areas when dealing with establishment waste:

Good waste management includes proper refuse and waste disposal.
Atlaspix/Shutterstock

- **Inside and outside waste containers.** Pests and wildlife will have no trouble finding a food source near soiled or overflowing containers. Containers should hold no more than they are built for. Check that container lids fit securely. Clean frequently, cleaning up any spillage on the ground.
- **Food preparation areas.** Clean up work areas regularly. Discard tainted food immediately. Keep worn food preparation clothing and reusable rags in sealed containers.
- **Bottles and cans.** Keep a tight lid on recyclable or returnable containers. Rinse out cans and bottles, as they provide a great home for pests. Remove cardboard boxes once they are emptied.

GENERAL HOUSEKEEPING

Other priority practices that eliminate pests from making the workplace a source of food and shelter include:

- **Good housekeeping in employee locker and break areas.** Food on worn clothing is a magnet for pests. Put stained or used clothing in sealed laundry hampers. Keep snack containers and drink cups picked up.

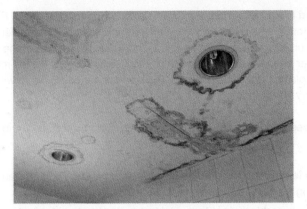

Repair ceiling leaks.
BOONROONG/Shutterstock

- **Practicing immediate spill and waste cleanup, both in food preparation and food serving areas.** In addition to pest prevention, debris on floors is a signal to customers as to the general cleanliness of the entire facility.
- **Frequent restroom checks.** Bathrooms are important to keep clean, especially during busy serving hours.
- **Rinsing and storing floor cleaning materials after use.** Empty mop buckets immediately after use. Rinse out mops with clean water. Dry and store cleaning tools after use.

PEST ENTRANCE

Pests can enter a facility through gaps, holes, or openings in the building. Examples of openings include cracks in the wall, a gap between walls and window frames, and open doors. People, supplies, and new equipment all have the ability to bring pests into the facility.

To prevent pests from entering the building:

- **Seal all wall openings.** Notice light or feel air coming through a wall. Use a sealer to correct openings.
- **Patch floor cracks and holes.** Look for hairline openings in the floor, especially around uneven areas. Use a patch that can both seal and withhold the activity of the area.
- **Repair ceiling leaks.** Follow a drip or water mark to its source—water can travel for some distance along rafters and trusses before becoming visible. If the leak is from an outside roof, have a professional inspect and maintain roofing materials. Contact building maintenance if the problem is coming from a different part of the facility.

To prevent pests from entering through doors and windows:

- **Maintain door seal.** Check for correct door fit. Maintain weather stripping for a tight seal on the door top and sides. Install a door sweep to seal the door bottom.
- **Post signs.** Signs can help to remind workers and customers to keep doors closed.
- **Consider an air curtain.** An air curtain creates a bug barrier by directing a flow of air away from the door, pushing flying bugs back and out of the facility.
- **Inspect window seals.** Install a sealant in gaps between doors and window frames and the walls.

Maintain door seals.

- **Install and maintain screens on doors and windows that open.** For doors and windows that are kept open for ventilation, the FDA Food Code specifies that the required screen mesh is 16 mesh to 1 inch/25.4 mm. The size of the mesh refers to the number of spaces that exist within a square inch of the woven material (such as wire, fiberglass, or plastic). A size of 16 wires per 1 inch/25.4 mm will protect the food facility from rodents and insects.

To protect against pests entering through plumbing and heating, ventilation, and air-conditioning (HVAC) systems:

- **Eliminate leaks and drips.** Pests thrive in moist areas. Fix leaky faucets, dripping or sweating pipes, and leaks in water or drainage piping.
- **Inspect dishwashing machines and hot water heaters for leaks.** Check fresh water pipes and drainpipes, especially if rubber or synthetic material is installed.
- **Guard floor drains.** Cover floor drains with a removable grate or strainer. Immediately clear plugged drains.
- **Seal gaps around pipe and duct work that passes through walls.** Select a sealant appropriate to the purpose—synthetic, metal, or concrete. The PCO can provide guidance on locations and types of materials to use.
- **Install screens over external ventilation vents and ducts.** As with screens for windows and doors used for ventilation, install 16 mesh screens. Clean often.

To protect against pests in food and food deliveries:

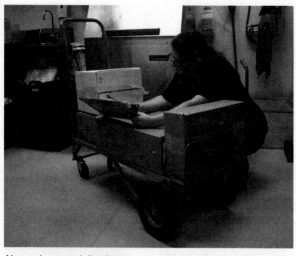

Always inspect deliveries.

- **Select a reliable supplier approved by a local regulatory authority.** Suppliers should be sensitive to food service pest control concerns.
- **Inquire into a supplier's pest control plan.** What do the suppliers do to ensure stock rotation? Are pesticides used in storage areas? How is spoilage handled? How are pests prevented from entering trucks or the building at the loading dock?
- **Inspect deliveries.** Inspect the truck for cleanliness. Open crates and cartons. Look for signs of pests, such as odor, droppings, or dead pests.

When inspecting and correcting building access issues, consult with the PCO or local health department. The PCO and local inspector should both have a wide knowledge of identification and correction techniques to support pest prevention efforts.

OUTSIDE DINING AREAS

Many of the same pest prevention and control practices that apply to indoor eating areas also apply to outside eating areas. However, outside dining areas provide their own unique pest control challenges. Unlike indoor areas, outdoor dining areas have no barriers, such as walls, doors, or windows, to keep pests out. Ants, bees, flies, and even birds have free access to these areas.

The challenge a food manager faces when planning for outdoor eating areas is to minimize pest incursions rather than achieve complete prevention. To decrease pest activity in outdoor dining areas:

- **Bus and clean tables immediately.** Remove used plates and glasses. Clean tabletops and inspect for spilled food or drink on floors and seats. Keep condiment containers clean and sealed.
- **Maintain landscaping.** Trees, bushes, and ground cover are all natural hiding places for pests. Keep grass mowed, bushes trimmed, and weeds under control. When possible, create a gravel or concrete space between the outside of the eating area and the landscaping or lawn.

Outdoor dining requires taking different measures to control pests.

Blend Images/SuperStock

- **Keep the immediate area dry and picked up.** Water attracts many pests, especially mosquitoes. Eliminate standing water. Pick up papers and other refuse.

- **Give special attention to outside dumpsters and garbage containers.** Relocate dumpsters as far away and as downwind as possible from the eating area. Periodically contact refuse companies to pressure-clean facility dumpsters. Keep all garbage cans clean and sealed tightly—place dirty tableware inside, behind a closed door.

- **Discourage customers from feeding wildlife.** Feeding wildlife creates an open-ended invitation for birds and other animals to find all their food at the facility. Install signs that explain the problem to customers.

- **Contact a PCO for additional help in minimizing the effect of pests in outside dining areas.** A PCO can remove nests and hives. Bug zappers, traps, and other pest control tools can be helpful in reducing the occurrence of pests. PCOs can provide valuable assistance in the proper placement and maintenance of this equipment.

Objective: Describe the proper use and storage of pesticides.

Pesticides are chemicals used to kill pests. Pesticides can cause serious illness and other injuries in humans, and must be handled properly.

Pesticides are packaged and distributed in many forms. Some examples include:

- **Bait, bait traps, glue boards, and tapes.** Baits and traps use a mixture of food to attract and pesticide to kill. Bait stations can permit the insects to ingest the poison and return to their colonies. Bait traps, glue boards, and tapes are designed to hold the pests. Bait stations are a popular, longer-term solution in eliminating rodents. Glue boards without pesticides are also used for rodent control, although they can be ineffective since rodents can escape.

- **Dusts and powders.** Dry mixtures can be used to get into cracks and crevices where insects hide.

- **Liquids.** Pesticides in concentrated form may require dilution before use. Some pesticides are distributed as ready-to-use liquids to get into cracks and crevices, similar to dusts and powders.

- **Sprays.** Contact, fogger, fumigant, and residual sprays are all options for pesticide use. Contact sprays are multiple-spray aerosol applicators that are sprayed directly onto pests. Foggers are complete release aerosol applicators used to clear a large area of an infestation. **Fumigation** methods require extended-use chemical compounds that vaporize over a long period of time. Residual sprays are also extended-use pesticides, and are applied to surfaces for a long-lasting effect. Sprays should never be used when food is out in the open.

Pesticides must be treated with caution. Many products can cause damage if used improperly. The following are questions to ask after ineffective or improper application of pesticides:

- **Unnecessary exposure.** Using non-chemical controls is the best protection from the unwanted effects of pesticides. Pesticides should only be used as a last resort. What other control options are available? How effective are they compared to the pesticide?

- **Failure to remedy infestation.** Was the pest correctly identified? Was the application too much or too little? Was there a need for a retreatment? Has the pest built up immunity to the pesticide?

- **Introduction of a hazard to employees and customers.** Were labels followed accurately? Was the pesticide for food service use? Was an outdoor-only pesticide used indoors?

- **Health hazard to a food service worker.** Did the label require use of personal protective equipment? Should rubber gloves and eye goggles have been worn during application? Should respiratory protection have been worn?

- **Food contamination.** Was a pesticide applied with food present? Were utensils removed and equipment covered? Were food preparation work areas cleaned after application? Was the area properly ventilated?

If a pesticide must be used, food service workers should follow these guidelines in selection and application:

- **Read and follow label directions.** Abuse of pesticides most commonly results from not reading label instructions. For additional information about a pesticide, obtain a Safety Data Sheet (SDS). An SDS is required for every pesticide and can be found online. Restricted-use pesticides can only be applied by a PCO.

- **Always use and store pesticides in original containers.** In addition to guaranteeing proper identification, this will ensure that the product label is readily available for reference or for other users.

- **Apply according to label directions.** Application directions are established based on the type of pest and threshold levels for effectiveness.
- **Wear personal protective equipment.** In addition to label information, the SDS has very specific directions for protecting the user, including what actions to take in case of contact, inhalation, or ingestion.

Safety Data Sheet

according to ANSI Z400.1 -2004 and 29 CFR 1910.1226

WEYLAND'S ANT KILLER 16 - UNSCENTED

Version *1.0*

Revision Date 04/11/2008

Print Date 03/04/2011

SDS Number 350000009774

---

1. PRODUCT AND COMPANY IDENTIFICATION
Product information

Trade name	: WEYLAND'S ANT KILLER 16 - UNSCENTED
Use of the Substance/Preparation	: Insecticide
Company	: Weyland Products, Inc. 1234 Random Street Anytown, AB 12345-1234
Emergency telephone	: 24 Hour Transport & Medical Emergency Phone (123) 555-2368 24 Hour International Emergency Phone (321) 555-2386

---

2. HAZARDS IDENTIFICATION

Emergency Overview

Appearance / Odor : clear / aerosol / Characteristic Odour

Emergency Overview

: Caution
CONTENTS UNDER PRESSURE. Do not puncture or incinerate. Do not store at temperatures above 120 Deg. F (50 Deg C), as container may burst. Keep away from heat, sparks and flame. Avoid contact with skin, eyes and clothing.

Potential Health Effects

Routes of exposure	: Eye, Skin, Inhalation, Ingestion.
Eyes	: May cause: Mild eye irritation
Skin	: May be harmful if absorbed through skin.
Inhalation	: May cause nose, throat, and lung irritation.
Ingestion	: May be harmful if swallowed. May cause: Abdominal discomfort.
Aggravated Medical Condition	: None known.

---

2. HAZARDS IDENTIFICATION

Chemical Name	CAS-No.	Weight %
Deionized Water	7732-18-5	50.00 - 60.00
DISTILLATES (PETROLEUM),	64742-47-8	10.00 - 30.00

1/8

A Safety Data Sheet (SDS) provides comprehensive information about the hazards of a chemical product, such as a pesticide. The law requires that chemical manufacturers, distributors, and importers provide SDSs for each hazardous chemical to communicate information about these hazards.

Improper storage of pesticides can create a health hazard, as well as compromise the effectiveness of the pesticide. Improper disposal can pose the same health hazards outside the facility and also harm the environment. The following are guidelines for storing and disposing of pesticides:

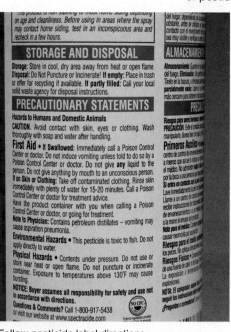

Follow pesticide label directions.

Science Source/Science Source

- **Follow all label and SDS directions for storage and disposal.** Contact a local PCO for assistance in compliance with all pesticide storage and disposal practices. Some state and local governments also have rules that may be stricter.
- **Buy only the amount of pesticides needed.** Purchase enough for a single infestation, or for a current season. It is better to dispose of an empty container than a full one.
- **Store within temperature guidelines.** Temperatures that are too high or too low can reduce a pesticide's effectiveness. Fire hazards can occur in higher temperatures; for example, aerosol cans stored above 120°F (49°C) can lead to a release and possible fire hazard.
- **Store in a sealed cabinet away from food products, food preparation materials, and other cleaning materials.** Include inspection criteria for cabinets on the cleaning schedule. Consider storage in an outside area, temperature permitting. Label as "Pesticide Storage."
- **Never dump unused pesticides into the sewer system.** Some water facilities are not equipped to filter out the chemicals. Follow the guidance of the PCO and label directions for safe disposal. Offer unused, labeled pesticides to another facility for their use. If the community has a waste disposal day, save them for drop off.

KEY TERMS

Fumigation A method of pest control that completely fills an area with smoke, gas, or vapor in order to kill vermin or insects.

Integrated Pest Management (IPM) An approach to pest control that uses a wide range of practices to prevent and solve pest problems in food facilities.

Pest An animal, bird, or insect capable of directly or indirectly contaminating food.

Pest control operators (PCO) Individuals licensed to control pests in the state in which they operate.

Pesticides Chemicals used to kill pests.

ASSESSMENT QUESTIONS

1. The primary goal in integrated pest management is:
   a. Calling the right PCO
   b. Ensuring staff knows the signs of pests
   c. Creating a thorough cleaning schedule
   d. Keeping pests out of the facility

2. Which of the following is a sign of rodent infestation?
   a. Cats hanging around outside the facility
   b. Shredded paper or cardboard piles
   c. Casings on the floor
   d. Strange, oily smell

3. What is the best resource for more information about a particular pesticide?
   a. A PCO
   b. FDA Food Code
   c. Back of the pesticide container
   d. Safety Data Sheets

4. What is one way to minimize pests in outdoor dining areas?
   a. Only serve food indoors.
   b. Maintain landscaping.
   c. Leave trash outside to deter pests.
   d. Set out food beyond the outdoor dining area to attract pests there instead.

5. Which type of pest could cause Salmonellosis?
   a. Cockroach
   b. Rodent
   c. Fly
   d. Both A and B

6. Which of the following is NOT a basic goal of most IPM plans?
   a. Applying pesticides whenever there are signs of pests
   b. Preventing breeding by removing any food or water sources that could be shelter areas to pests
   c. Working with a licensed pest control operator (PCO) to control any infestations that occur
   d. Preventing pest entrance to the facility

7. Pests are most attracted to:
   a. Clean kitchens
   b. Empty dumpsters
   c. Light areas
   d. Moist areas

8. Upon noticing signs of a pest infestation, what should you do first?
   a. Apply a pesticide in that area.
   b. Take a photo to show the PCO.
   c. Call your PCO and describe what you found.
   d. Post a notice to warn customers.

9. Which of the following is NOT an example of a pesticide?
   a. Sprays
   b. Liquids
   c. Bleach
   d. Glue traps

10. Why should you never dump unused pesticides into the sewer systems?
    a. They might clog your pipes.
    b. Some water facilities are not equipped to filter out the chemicals.
    c. Disposing of unused pesticides is a waste of money.
    d. A license is required to dispose of pesticides.

QUESTIONS FOR DISCUSSION

1. List some common pests found in the food industry.

2. What is an IPM team? What are some ways that the PCO supports the IPM team?

3. Describe three ways to prevent pests from entering the building of your restaurant.

4. Improper use of pesticides by food service workers can have very serious consequences. What are some of those potential consequences?

EMPLOYEE TRAINING

E mployees need to understand the importance of following food safety principles immediately after starting their jobs. New staff members will not be familiar with a company's food safety procedures without the proper training. It is the responsibility of the food safety manager to provide thorough training for new employees and frequent refresher training sessions for all employees. Proper training should include safety, personal hygiene, safe food preparation, cleaning, sanitizing, and pest recognition and prevention.

After reading this chapter, you should be able to:

- Describe the relationship between personal hygiene and food safety hazards.

- Explain why hand washing is important for food handlers.

- Explain the importance of wearing gloves.

- Discuss the importance of a personal hygiene policy in a food facility.

- Demonstrate the importance of communication in the workplace.

- Create a successful training schedule.

Proper clothing
Phovoir/Shutterstock

Learning Objective: Describe the relationship between personal hygiene and food safety hazards.

Apron
michaeljung/Shutterstock

It is imperative that employees who handle food maintain a high level of **personal hygiene**. Employees should be in the practice of showering or bathing on a daily basis, and they should make a habit of frequent, proper hand washing.

CLOTHING

Food handlers must always wear clean and washable protective clothing. Protective garments should be appropriate for the work being carried out. Protective garments should also completely cover ordinary clothing.

It is important for employees to dress from the top down to avoid contaminating protective clothing with bacteria from hair and work shoes. Be sure employees follow the following steps in order:

- Put on the hairnet or the hat.
- Dress in the other pieces of clothing.
- Put on shoes.

Remember, the main purpose of protective clothing is to protect food from contamination. If possible, food handlers should change into work clothes once they have entered the work facility.

APRONS

Food handlers should begin the workday with a fresh, clean apron and should remove aprons when leaving food preparation areas. There should be a designated area to hang or store aprons any time a food handler has to remove an apron and take out trash or visit the toilet. If an apron becomes soiled, it should be changed immediately.

SHOES

Suitable footwear should be worn to prevent slipping and to protect the feet. Appropriate footwear includes clean, closed-toe shoes with a low heel and nonskid soles.

JEWELRY, PERFUME, AND NAILS

Other potential sources of contamination by food handlers are jewelry, perfume, and long or fake fingernails. Food handlers should not wear earrings, watches, or rings, as they hide dirt and bacteria. Stones and small pieces of metal in jewelry may contaminate food, as can nail polish or false nails. The only jewelry permitted is a plain wedding band. Strong-smelling perfume or aftershave can taint food taste and smell and should be avoided by food handlers.

SMOKING

Smoking in food preparation areas or while handling food is illegal. Food handlers who smoke should do so only in designated areas.

Nonskid soles
Stocksnapper/Shutterstock

Some reasons food handlers must not smoke in food preparation areas include:

- People touch their lips and can transfer bacteria to food from their mouth.
- Cigarettes contaminated with saliva may be placed on work surfaces.
- Smoking encourages coughing.
- Cigarette butts and ash may land on and contaminate food.

In many areas of the United States, recent legislation has banned smoking from taking place anywhere inside the facility. Every manager should designate an area outside the premises where smoking is permitted.

POLICIES

It is always a good practice to check with local regulatory agencies regarding the requirements for proper work attire and policies on smoking, eating, and drinking. Make sure the establishment's handbook has written policies so that current and potential employees are aware of what is and what will be expected of them when reporting to work.

SETTING PERSONAL HYGIENE STANDARDS

All people, especially food handlers, are potentially the greatest hazard in a food facility. They are sources of physical, chemical, and biological hazards, and they can cause cross-contamination as a result of poor hygiene practices.

Good personal hygiene habits reduce or eliminate the main food safety hazards associated with people. A **food safety hazard** is a biological, chemical, or physical agent in food, or a condition of food, with the potential to cause harm (that is, an adverse health effect) to the consumer. The basic personal hygiene program should expect that employees:

It is essential to practice good personal hygiene.

- Practice proper hand-washing techniques, and employ proper glove use when applicable.
- Maintain a high level of personal cleanliness.
- Wear proper work attire.

PREVENTING CONTAMINATION

Contamination can be controlled or eliminated in many ways. The principal hazards associated with the human body are bacterial. Proper hand washing will greatly reduce these hazards. *Staphylococcus aureus* is the most common bacteria associated with humans. *S. aureus* is often present in boils, in skin infections and cuts, on hands, in the nose, mouth, and ears, and on hair.

When it comes to boils, skin infections, and cuts, make sure that:

- Food handlers cover all wounds with a waterproof dressing, preferably blue (so that the dressing can easily be seen if it falls into food)
- Staff members who have boils, **lesions**, or infections be excluded from handling time/temperature control for safety foods

Keep hands clean at all times.

HANDS

Hands often act as a vehicle to transfer bacteria to food, since they come into direct contact with food during preparation.

When it comes to hands, employees must:

- Keep hands clean at all times
- Keep nails short and clean
- Not use false nails or nail polish

NOSE, MOUTH, AND EARS

When it comes to the nose, mouth, and ears, employees must adhere to specific rules. The food manager must ensure that employees do not:

- Cough or sneeze over food or equipment
- Bite nails, blow noses, or scratch ears or skin
- Lick fingers, blow onto food or equipment, or spit
- Snack or chew gum in food-handling areas
- Use a finger to taste the food (instead of using a clean spoon)

HAIR

Many germs are found in human hair. When it comes to hair, insist that employees:

- Wear hairnets, protective hats, or even hairnets under protective hats when possible
- Do not touch or comb their hair while at work
- Do not scratch their heads or beards
- Keep their hair clean and short or keep longer hair tied back

Do not blow on food to cool it

Karramba Production/Shutterstock

Learning Objective: Explain why hand washing is important for food handlers.

Hand washing is one of the most important actions that can be taken to prevent the spread of foodborne illnesses.

MANAGER RESPONSIBILITY

It is the food manager's responsibility to train employees in the proper hand-washing technique. The hand-washing method must be monitored and put into practice. A good manager will lead by example and practice proper hand washing at all times. To ensure that staff adhere to these good practices, place posters in the food area, demonstrate regularly, and remind employees of the proper technique.

WHY HAND WASHING IS IMPORTANT

The main objective of washing hands is to reduce the number of pathogens on hands to a safe level. An employee working with salads who carries *Shigella* bacteria and does not wash his or her hands can pass the bacteria to customers through the salad.

A single-wash procedure is normally sufficient. However, activities likely to result in a large number of pathogens on the hands should be followed by a double-wash instead. If the hands are likely to be heavily contaminated, for example, after going to the toilet, changing a dressing, or cleaning up feces or vomit, then a nailbrush should be used before the normal wash. The double-wash procedure includes an extra, initial stage to brush the fingernails and fingertips under running water, using liquid soap on a nailbrush. This additional step will decrease the chances of contamination.

Bacteria is easily spread through bare-hand contact.
Lightspring/Shutterstock

WHEN TO WASH HANDS

Food handlers must wash their hands before starting work and should wash them regularly throughout the workday, especially following certain activities. Hands must always be washed after using the toilet, without exception.

Food employees must wash their hands before:

- Entering a food preparation or serving area
- Handling time/temperature control for safety food
- Handling ready-to-eat food

Food employees must wash their hands after:

- Clearing tables or busing dirty dishes
- Touching or taking out the trash
- Dealing with an ill customer or coworker
- Putting on or changing a dressing covering boils, skin infections, or cuts
- Cleaning animal feces, handling boxes contaminated with bird droppings, or handling a baby's diaper
- Handling hazardous chemicals
- Touching the hair or face
- Eating or smoking
- Coughing, sneezing, or blowing the nose
- Handling money
- Handling raw food

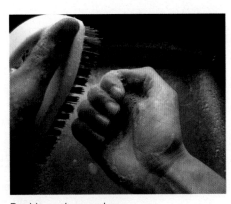

Double-wash procedure
Muriel Lasure/Shutterstock

HOW TO WASH HANDS

To reduce bacteria to a safe level through hand washing, it is essential that the correct procedure be followed. Proper hand washing should take about 20 seconds and follow the method outlined here:

1. Wet hands and exposed portions of the arms with clean, hot, running water. The temperature should be at least 100°F (38°C).

2. Apply a liquid soap.

3. Rub the hands together vigorously, cleaning all parts of the hands and arms, especially the fingertips and around the nails. Do this for 10 to 15 seconds.

4. Rinse hands and arms.

5. Completely dry hands and arms using a single-use paper towel or an air dryer.

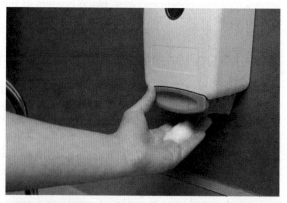

How to wash hands

WHERE TO WASH HANDS

In order to avoid contamination, hand washing must take place in a hand washing–specific basin, such as a dedicated hand-washing sink or an approved automated hand-washing facility. Hand washing should not occur in sinks used for preparing food or in a service sink.

BARE-HAND CONTACT

The current FDA Food Code prohibits bare-hand contact with ready-to-eat (RTE) food; that is, food that is going directly to the consumer without further cooking or preparation on food service premises. Employees may not handle RTE foods with bare hands in a facility that serves a highly susceptible population (such as preschool-aged children, the elderly, or the immunocompromised). This does not pertain to ready-to-eat food that is being added as an ingredient to food that will be appropriately cooked. If contact with RTE food is necessary, have the appropriate utensils available to employees, such as:

Bare-hand contact
Kzenon/Shutterstock

- Deli tissue
- Spatulas
- Tongs
- Scoops
- Single-use gloves
- Dispensing equipment

In some cases, approval to touch RTE food with bare hands may be obtained by contacting the proper regulatory authority.

Employees must sign that they have received the following trainings if they are ever to make bare-hand contact with food:

- Risks associated with contacting specific ready-to-eat food items with their bare hands
- Proper hand-washing methods, including when and where to wash hands
- Proper fingernail maintenance, including the proper use of a nailbrush
- Prohibitions against jewelry, such as rings, where bacteria and germs can evade the effects of hand washing

Employees should be aware of local health policies, such as what to do in cases of illness, when illness will exclude employees from bare-hand contact or the workplace altogether, and when illness might result in restrictions to bare-hand food contact or food contact in general.

In jurisdictions that have not yet adopted the 2005 FDA Food Code in its entirety and still allow bare-hand contact with RTE food, the food service establishment must utilize the following safeguards. Please note that these safeguards must be utilized in addition to, not instead of, regular and proper hand washing.

- Double hand washing before contact
- Use of nailbrushes during hand washing
- Use of hand sanitizer following proper hand washing

The final step required for bare-hand contact of ready-to-eat foods is documentation of corrective actions taken when steps one and/or two are not followed. This documentation must include a plan for implementing corrective actions. It may be included as a part of the food service establishment's HACCP log (described in Chapter 7) or as a stand-alone document, but it must be in written format.

A few states still allow bare-hand contact with ready-to-eat food. As an extra precaution, however, the FDA Food Code recommends that all RTE foods be touched only with gloved hands.

Learning Objective: Explain the importance of wearing gloves.

In order to maintain a clean and safe environment, employees should wear gloves to inhibit the spread of bacteria.

When used properly, gloves can aid in the service of safe food by acting as an added layer of protection between hands and food. When customers see employees wearing gloves, it tells them that the business is committed to serving safe, quality food and to protecting customer and employee health.

When it comes to purchasing gloves for food handlers, there are several factors to be aware of:

- Gloves used in food service facilities should be single-use gloves only. Do not wash and reuse gloves.
- Buy gloves in a variety of different sizes. If gloves are too large, employees will have a hard time keeping them on their hands. If the gloves are too small, they will rip or tear.
- Do not use latex gloves. It is common for people to be sensitive to latex or to develop allergies due to the extended use of latex gloves. Alternative materials include polyvinyl, nitrile, chloroprene, and polyethylene.
- Match the proper type of glove to the appropriate task. Buy looser-fitting, less expensive gloves for times when frequent changing is necessary. Buy more durable, form-fitting gloves that cost a little more for repetitive tasks that require less frequent changing.

The use of gloves does not replace proper hand washing. In fact, before putting on gloves, food handlers must always wash and dry hands using the proper hand-washing technique. Because gloves act as a second skin, they can be contaminated in the same way that hands can. Gloves should be changed after touching anything that may be a source of contamination.

Food employees must change their gloves:

- When changing tasks
- After touching raw meat
- Before handling cooked or ready-to-eat food
- After touching the mouth when sneezing or coughing
- After touching face or hair
- When they become soiled or torn
- After four hours

Gloves should never be worn for more than four hours, as perspiration and bacteria can build up under the gloves. After four hours, even if the gloves haven't been contaminated, employees should wash their hands and replace the gloves.

Gloves act as an added layer of protection between hands and food.
khz/Shutterstock

LESSON 4 | EMPLOYEE HEALTH

Learning Objective: Discuss the importance of a personal hygiene policy in a food facility.

PERSONAL HYGIENE POLICY

The personal hygiene policy should be clear and easy to follow. The policy must outline the actions for management and employees to take when they are sick. It is important that sick employees notify management before coming to work, or if they become ill while on the job. It is also vital that managers remain approachable so that employees feel comfortable communicating to supervisors when they are ill.

RESTRICTING EMPLOYEES FROM WORK

Food handlers are particularly hazardous when they are ill. Everyone on staff must know it is mandatory that they report any illnesses promptly. Managers will have times when **restriction** of an employee from working with food, from working in food contact areas, or from being present in the establishment is the only solution. For example, an employee must be restricted from working with food or in food contact areas if he or she has a sore throat and fever. If serving a highly susceptible population (HSP) (such as the elderly, the very young, people who have compromised immune systems, pregnant women, or allergen-sensitive people), certain employees must be excluded from the establishment. **Exclusion** from any establishment is necessary for an employee who is vomiting, or has diarrhea or jaundice.

Employees must notify management if they are experiencing any of the following symptoms:

- Vomiting
- Diarrhea
- Jaundice
- Sore throat with fever
- A wound or lesion, such as a boil or infected wound, that is open or draining and cannot be protected by a waterproof, properly fitting cover or bandage

It is management's responsibility to notify regulatory agencies when an employee's illness is serious enough. When an employee is diagnosed with any of the following illnesses, he or she must be restricted from working with food in any context, and excluded from the establishment if a highly susceptible population (HSP) is to be served:

- Norovirus
- Hepatitis A virus
- *Shigella* spp.
- **Shiga toxin–producing *E. coli***
- *Salmonella* typhi
- Non-typhoidal *Salmonella*

Good hygiene will keep customers and employees healthy
SK Design/Shutterstock

An employee must notify management if he or she has had an illness due to *Salmonella* typhi within the past three months and has not received antibiotic therapy.

If an employee has been excluded from work due to an illness or disease that requires regulatory notification, the food manager must work with that employee's health care practitioner and/or the regulatory agency to determine when the employee may return to work. No matter the cause, whenever an employee previously restricted from work returns to work, it is critical that good hygiene, particularly hand washing, is observed at all times.

Sometimes food handlers may be healthy carriers of an illness. Without being sick, one can still spread disease. It is essential that management and staff observe the highest standards of personal hygiene at all times, regardless of wellness. Food managers should consider creating an incentive program that encourages and/or assists food employees to refrain from working when ill or contagious. This could include paid time off for documented illness or reassignment in duties without impacting pay.

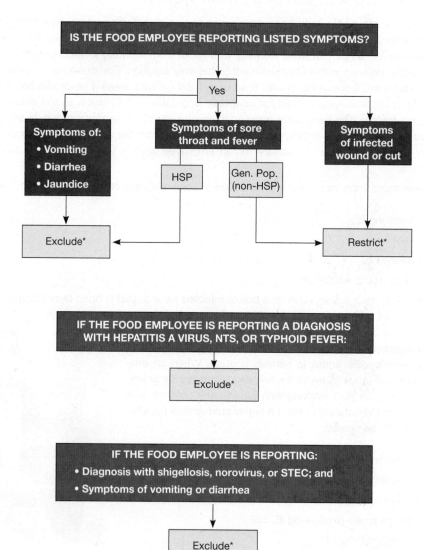

Exclusion and Restriction Charts from Annex 3, FDA Food Code

Key:

HSP = Highly Susceptible Population

HAV = Hepatatis A Virus

NTS = Nontyphoidal Salmonella

STEC = Shiga toxin-producing
            Escherichia coli

* Refer to the Exclusion and Restriction Charts from Annex 3, FDA Food Code, Lists I and II

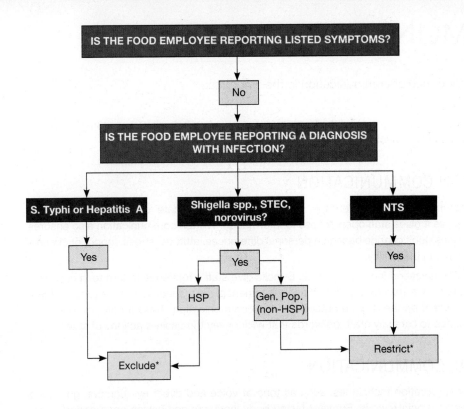

IS THE FOOD EMPLOYEE REPORTING LISTED SYMPTOMS?

No

IS THE FOOD EMPLOYEE REPORTING A DIAGNOSIS WITH INFECTION?

| S. Typhi or Hepatitis A | Shigella spp., STEC, norovirus? | NTS |

Yes — HSP / Gen. Pop. (non-HSP) — Yes

Exclude*

Restrict*

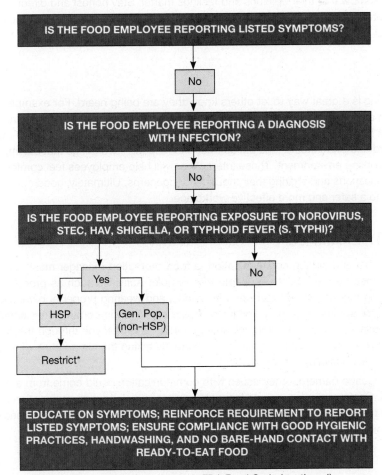

IS THE FOOD EMPLOYEE REPORTING LISTED SYMPTOMS?

No

IS THE FOOD EMPLOYEE REPORTING A DIAGNOSIS WITH INFECTION?

No

IS THE FOOD EMPLOYEE REPORTING EXPOSURE TO NOROVIRUS, STEC, HAV, SHIGELLA, OR TYPHOID FEVER (S. TYPHI)?

Yes   No

HSP   Gen. Pop. (non-HSP)

Restrict*

EDUCATE ON SYMPTOMS; REINFORCE REQUIREMENT TO REPORT LISTED SYMPTOMS; ENSURE COMPLIANCE WITH GOOD HYGIENIC PRACTICES, HANDWASHING, AND NO BARE-HAND CONTACT WITH READY-TO-EAT FOOD

Exclusion and Restriction Charts from Annex 3, FDA Food Code (*continued*)

Key:

HSP = Highly Susceptible Population

HAV = Hepatatis A Virus

NTS = Nontyphoidal Salmonella

STEC = Shiga toxin-producing
Escherichia coli

* Refer to the Exclusion and Restriction Charts from Annex 3, FDA Food Code, Lists I and II

Learning Objective: Demonstrate the importance of communication in the workplace.

Communication is key to a good working environment in any industry. Clear, complete directions up front help to avoid misunderstandings and problems in the future.

WRITTEN COMMUNICATION

Written communication is often the best way to communicate company policies and procedures, as it gives staff open access to information. Written communication also ensures that guidelines don't change based on personal differences, attitude, stress, mood, or circumstantial events in the facility.

Written communication includes e-mails. E-mails give staff a reference to turn to in case of a disagreement. An e-mail or any written announcement gives management time to really think about and reflect on the desired outcome of the communication. This can provide management a chance to carefully draft the words that will convey the desired actions of staff.

VERBAL COMMUNICATION

Verbal communication techniques, such as tone of voice and direct eye contact, go a long way in getting a point across. Forward-facing body language and asking open-ended questions can let staff know that their opinions and feelings matter. Stay honest and direct, allowing those receiving the communication to be heard as well. It is often the case that the most important part of communicating is listening.

LISTENING

Reflective listening is a great way to let others know they are being heard. For example, use comments such as, "It sounds like you are asking about [xyz]. Is that correct?" If a manager is able to reflect back on staff comments later in a conversation, the employee will know he or she was heard and understood. This builds trust and respect—two things that are imperative to a good working environment. These interactions will help employees feel comfortable asking difficult questions and sharing their thoughts or concerns. Ultimately, good communication can lead to a safer and more effective workplace.

LEADING BY EXAMPLE

In addition to verbal and written communication, a food protection manager must communicate and demonstrate food safety concepts by example. Activities such as proper hand washing, sanitizing thermometers before probing foods, and wearing proper apparel should be used to reinforce spoken or written directions. This is especially important when language barriers might pose a difficulty in communication. Leading an employee through the physical steps of washing and sanitizing equipment, or correctly taking the temperature of a food, remains a crucial responsibility of management.

In addition to language barriers, other issues with communication could come from a difference in cultures. Some examples include language styles, body language assumptions, use of emotion, and timeliness. Management should facilitate a workplace where employees can communicate comfortably.

Learning Objective: Create a successful training schedule.

Employee training enables food workers to acquire the capabilities they need to perform their jobs correctly. Training is linked to both employee performance and retention. Food managers can provide training every day by following these four main stages:

- Motivating staff
- Teaching new information and processes
- Supervising tasks and procedures
- Testing staff understanding through questions and observation

METHODS

There are many different training methods that food managers can use to convey knowledge and skills. It is beneficial to use a variety of methods to ensure that training is effective, because each employee learns differently.

When determining what delivery methods will work best, recognize that training must be delivered in a manner that is appropriate to the training content, workplace, and audience.

For example, it would not be appropriate to give examples throughout the training that are not relevant to the staff member's workplace, or to give in-depth **microbiological** explanations to front-line staff members.

When planning training and methods of delivery, consider:

- The employee's learning style. For example, are they very hands-on, do they like to go through written documents on their own, or do they find working at their own pace on the computer less threatening?
- Whether there are any barriers to learning. This would include such things as language, attitude, learning difficulties, or any disabilities.

VALUE PROPOSITION

Employees need to see the value in training. Make it clear how the training will make their jobs easier, help them work more effectively, and help them become more successful.

- **Offer hands-on activities.** Devote about two-thirds of training time to hands-on activities that allow staff members to practice their new skills.
- **Give feedback.** Remember to keep feedback immediate, specific, and positive, and offer it for both correct and incorrect behaviors.
- **Keep each training session short.** Break sessions down into small, manageable amounts of time and reasonable chunks of information. Short sessions of no more than 45 minutes are always best to ensure retention.

To ensure that training principles are reinforced in the workplace, food managers must:

- Have clearly defined objectives that can be measured
- Ensure that the training supports the identified objectives
- Evaluate employees after completion of the training to make sure the objectives have been met

Offer hands-on activities that allow staff to practice.

Minerva Studio/Shutterstock

- Foster a work environment that allows and encourages employees to put their training into practice
- Lead by example and reinforce the company's commitment to food safety

DOCUMENTATION

A training program isn't complete unless it is documented. Food managers must keep records of all food safety training conducted and employee training records. These records are beneficial for reviewing and evaluating an employee's job performance, and they are also good to have for legal purposes.

REFRESHERS

Refresher training is needed to ensure that all—not just new—employees stay up to date on all aspects of food safety. Refresher training should be carried out as needed and whenever required. For example, after new equipment is installed or when introduction of new legislation is occurring, additional training may be mandatory. Training should be done at regular intervals, as dictated by the company's policies. It can also be dictated by circumstances such as an incident or complaint, daily checks, or training reviews. Use refresher training to ensure staff's competency and to ensure that staff's knowledge and skills are current.

KEY TERMS

Exclusion   Requiring a worker to leave the food establishment as a result of specific illnesses, symptoms, or exposure to certain diseases.

Food safety hazard   A biological, chemical, or physical agent in food, or a condition of food, with the potential to cause harm (that is, an adverse health effect) to the consumer.

Hand washing   The process of cleansing the hands with soap and water to thoroughly remove soil and/or microorganisms. Food workers must clean their hands up to their elbows.

Lesion   A skin injury, usually caused by disease or trauma.

Microbiological   Of the branch of biology dealing with the structure, function, uses, and modes of the existence of micro-organisms.

Personal hygiene   Standards of personal cleanliness habits, including keeping hands, hair, and body clean and wearing clean clothing in the food establishment.

Restriction   Preventing a worker with certain illnesses or symptoms from working with food or in food contact areas.

Shiga toxin–producing E. coli   Also known as STEC, an infection-causing bacteria, containing the strain 0157:H7, found in ground meats, unpasteurized raw milk, and contaminated produce.

ASSESSMENT QUESTIONS

1. A food handler should wash hands before and after:
   a. Using the restroom
   b. Taking out the garbage
   c. Preparing raw hamburger
   d. Putting on an apron

2. What is the most important reason for promoting good personal hygiene?
   a. To increase morale in the workplace
   b. To remove unwanted odors from the facility
   c. To reduce contamination hazards associated with people
   d. To quiet the health inspector

3. What is the most common bacteria associated with people?
   a. *Staphylococcus aureus*
   b. *Salmonella*
   c. *Clostridium botulinum*
   d. *Clostridium perfringens*

4. What is the most effective way to communicate policies?
   a. Post them on the Internet
   b. Verbally remind your employees
   c. Recite them to your customers
   d. Write them down and share them with employees

5. What needs to be done when a food service employee complains of having a sore throat and a fever?
   a. Exclude the employee from the facility
   b. Report the illness to the local health agency
   c. Call the employee's doctor
   d. Restrict the employee from working with food

6. A food handler who cuts chicken for an entire shift should:
   a. Change gloves after an hour due to the amount of pathogens involved with chicken
   b. Change gloves at least every four hours
   c. Wear the same gloves for the entire shift
   d. Wash and sanitize the gloves to save money for the business

7. The four stages of training are:
   a. Research, implement, deliver, review
   b. Motivate, teach, supervise, test
   c. Initiate, demonstrate, review, test
   d. Introduce, convey, supervise, correct

8. Which of the following is NOT considered to be appropriate footwear for a food employee?
   a. Open-toed shoes
   b. Closed-toed shoes
   c. Low-heeled shoes
   d. Shoes with nonskid soles

9. In hand washing, food handlers must scrub their hands and arms for:
   a. 5 to 10 seconds
   b. 10 to 15 seconds
   c. 15 to 20 seconds
   d. 20 to 25 seconds

10. An employee must notify management if he or she has had an illness due to *Salmonella* typhi within the past:
    a. Three years
    b. Six months
    c. Three months
    d. Six years

QUESTIONS FOR DISCUSSION

1. Explain the importance of employee hygiene, with specific attention paid to clothing, footwear, and jewelry.

2. What are some specific instances of when hands must be washed in a food service environment?

3. List three activities that would require food employees to change their gloves.

4. When must an employee be restricted from working with food in any context, and excluded from the establishment if a highly susceptible population is to be served? List specific examples.

5. What are some forms of workplace communication? Which do you believe is the most effective?

6. Imagine that you are the new manager of a popular burger restaurant, with 30 employees reporting to you. How will you train them? Outline your training plan.

FACILITIES AND EQUIPMENT

F ood safety is directly impacted by the design of a food establishment, the flow of food through the facility, and the type of food storing and processing equipment used. High-quality food contact materials and regular maintenance of the equipment can help protect consumers from hazards associated with food. The environment in which food is produced is just as important as the quality of the food itself.

After reading this chapter, you should be able to:

- Explain how the design of a food facility can reduce cross-contamination.
- Describe how the use of certain food contact materials can cause contamination.
- Explain the importance of cleaning and sanitizing the food service facility.
- Describe the various washing facilities found in a food establishment.
- Explain the importance of safe drinking water in a food service facility.

Proper kitchen design is essential to avoiding hazards in the kitchen.
Kris Vandereycken, 2010/Shutterstock

Learning Objective: Explain how the design of a food facility can reduce cross-contamination.

Good design and regular maintenance of food facilities are essential to avoid hazards such as the contamination of food and multiplication of bacteria. The food premises must:

- Be large enough to accommodate all essential equipment
- Have separate areas for the storage and preparation of raw foods and ready-to-eat foods
- Have appropriate storage space for time/temperature control for safety (TCS) foods
- Ensure a continuous workflow from delivery to service
- Ensure the separation of food and waste

The food facilities must be kept clean and in good repair, or all design benefits will be lost.

The most important design principle in the food facility is to minimize the distance traveled by food and food handlers. The best facility design features are linear or continuous, progressing from raw material to finished product. This is vital to prevent contamination and cross-contamination, as it ensures minimal or no contact between raw foods and ready-to-eat foods.

An example of a food facility with a poor layout is one where raw food and ready-to-eat food are prepared together, which can lead to cross-contamination. A poorly constructed food facility is one built with materials that allow for condensation, or one that has open joints or gaps where dirt and bacteria can collect. An example of a poorly situated food facility is one that is located in an area prone to flooding, as this could result in sewage contamination. A facility located near a factory could result in chemical contamination.

FLOORS

Flooring materials must be selected with health and safety requirements in mind. All flooring should be sturdy and easy to clean. It is also important to select flooring that will withstand heavy use and to use anti-slip floors in areas when necessary. Anti-slip flooring should be used only in high-traffic areas.

Nonporous, resilient flooring

Areas of the food facility that need to use some kind of nonporous, resilient flooring include: walk-in refrigerators, dishwashing areas, restrooms, garbage storage spaces, and places where the floor may be subject to moisture, flushing, or spray cleaning.

Nonporous, resilient flooring is typically made of rubber or vinyl tile. It withstands shock, and it is fairly inexpensive, durable, easy to clean, grease resistant, and easily repaired or replaced. The only problem with nonporous flooring is that it can be damaged easily from sharp objects.

Hard-surface flooring

Another type of flooring is hard-surface flooring. Hard-surface flooring is commonly made of ceramic tile, brick, marble, or hardwood. Like nonporous, resilient flooring, it is also nonabsorbent and durable; however, this type of flooring is more expensive and not as easy to clean. Objects dropped on hard-surface flooring may be likely to break. Hard-surface flooring can be slippery, and it does not absorb sound well. Hard-surface flooring is a good choice to use in restrooms and other high-traffic areas such as entrances and lobbies.

Carpet

The FDA Food Code states that carpeting may not be used in walk-in refrigerators, dishwashing areas, restrooms, garbage storage areas, or any other places where the floor is subject to moisture, flushing, or spray cleaning. Carpeted areas are best used in dining areas, and should be vacuumed daily or more often if necessary.

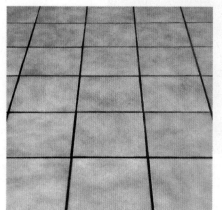

Hard-surface flooring
designelements/Shutterstock

Nonslip flooring

Nonslip flooring should be used in high-traffic areas—especially the kitchen. It is acceptable to use rubber mats in areas where there is likely to be standing water, such as the dishwashing area. These mats should be picked up off the floor and cleaned frequently. The floor under the mats should be mopped or scrubbed before the mats are replaced.

WALLS AND CEILINGS

Food facility walls should be sealed, sturdy, and easy to clean. Ceilings should be covered, and joists and rafters must not be exposed to moisture. Any utility service lines or pipes should not be unnecessarily exposed. Any fixtures attached to the walls in food areas—such as light fixtures, vent covers, and wall fans—must be easy to clean.

GOOD LIGHTING

Food facilities must have adequate lighting so that food handlers can carry out their tasks safely and correctly as well as identify hazards. Good lighting helps ensure effective cleaning and sanitizing. Fluorescent tubes fitted with protective sleeves or diffusers and shatterproof bulbs and guards are required where food could be exposed to risk from broken glass. Facility lighting has the added benefit of discouraging pests from entering the food facilities.

The FDA Food Code specifies different intensity levels for lighting, measured in lux or foot-candles, based on the function of that area:

- 108 lux (10 foot-candles): dry storage areas, walk-in refrigerators, and freezers
- 215 lux (20 foot-candles): buffets, bars, reach-in and under-the-counter refrigerators, hand-washing and dishwashing stations, equipment storage areas, restrooms
- 540 lux (50 foot-candles): food preparation surfaces with knives, slicers, grinders, and other utensils

Covered ceilings and good lighting can help identify hazards.
Suprun Vitaly/Shutterstock

Restaurant ventilation systems improve indoor air quality.
Chan Thitidechanant/Shutterstock

VENTILATION

In general, ventilation should be sufficient to keep facility temperatures and humidity at appropriate levels and maintain uncontaminated indoor air. A proper ventilation system helps remove vapors, odors, and fumes that can cause contamination. Ventilation should flow from the interior of the kitchen or preparation areas to the outside of the building. This is called positive airflow and helps to prevent pest entrance to the facility. Positive airflow ensures that smoke and grease are always exiting the building rather than recirculating. If the food facility has proper ventilation, there will be minimal grease buildup on walls and ceilings.

Ventilation systems must be designed and installed according to legal specifications. Ventilating systems, heating systems, and air-conditioning systems must allow for the intake of make-up air at installation and use exhaust vents that do not cause contamination. Make-up air vents cannot be added after ventilation or building temperature systems are installed. Make-up air is extra ventilation that creates a more healthful interior environment and ensures that plenty of fresh air is entering the facility. Make-up air intakes must be screened and filtered to prevent dust, dirt, insects, and other contaminating material from entering the facility. Management should evaluate ventilation systems to be sure they meet high standards.

The ventilation system must be configured so that hoods, ductwork, and fans don't drip onto food or equipment. All hoods should be tested prior to use, fit securely, and be easy to remove and clean.

The purpose of an exhaust hood is to provide a way to collect all of the grease produced from cooking. It also serves as a means of removing heat, smoke, and odors from the cooking area. For the hood to do its job, there must be enough air movement to catch grease particles and cooking odors and pull them up to the grease extractors. This keeps the grease particles from clinging to nearby surfaces.

The ventilation equipment, including hoods and ductwork, should be cleaned regularly. Check with local regulatory agencies to ensure that the ventilation system in the food facility meets all requirements and standards.

WASTE

Careful consideration should be given to the storage and disposal of refuse, waste food, and unfit food in the facility. Some measures that can prevent contamination from waste include the following:

- Require suitable receptacles both inside and outside the facility.
- Plastic garbage bags inside plastic or rubber bins provide an extra barrier for handling garbage inside the facility.

- Foot-operated bins are recommended for internal use, as bin lids are common vehicles of contamination.
- Empty bins frequently into the external waste container and never allow overflow.
- When taking the inside garbage to the outside receptacles, do not carry the bags over food or food preparation surfaces, in order to prevent potential contamination.
- Outside receptacles should be stored on concrete or asphalt, not dirt.
- Outdoor garbage receptacles are typically large metal garbage cans or dumpsters.
- External receptacles must be waterproof and leakproof, have secure lids, be made of metal or heavy-duty plastic, and be easy to clean.
- The outside refuse area must be kept clean.
- The area around the external receptacles should be hosed down regularly, and drains should be kept clean to prevent odors and deter pests.
- Recyclables need to be stored in a separate area away from food.

Suitable receptacles should be provided both inside and outside the premises.
vichie81/Shutterstock

Whether designing new facilities or remodeling existing facilities, the food manager should check with local regulatory agencies about approval of design plans. Even if approval is not required, a plan review will ensure compliance with any existing regulations. If the review uncovers any flaws, resolving the issues before construction begins will save time and money in the long run. The Americans with Disabilities Act **(ADA)** outlines requirements for premises that apply to employees and guests with disabilities. ADA accessibility guidelines are an important reference for food managers and contractors alike. Public businesses must modify policies and practices that discriminate against people with disabilities; remove barriers; provide services when needed; and comply with accessible design standards when constructing or altering facilities.

After construction, the food facility management must obtain an operating permit. Before the permit is granted, the facility may have to undergo inspection to ensure that required design specifications are met. Once the facility has passed inspection, it will be ready to open.

Restaurants must be accessible to people with disabilities.
CREATISTA/Shutterstock

Learning Objective: Describe how the use of certain food contact materials can cause contamination.

In order to prevent contamination, food managers must continually ensure that kitchen equipment is clean and well maintained. Standards that management should maintain are as follows:

- Equipment must be designed for kitchen use and made of materials that will not collect dirt and bacteria.

- Food equipment that is made from inappropriate materials or that is cracked, chipped, broken, worn, or badly designed is a haven for dirt and bacteria. It can also become a source of physical or chemical contamination.

- Equipment that comes into contact with food should be smooth, waterproof, nontoxic, non-flaking, non-tainting, resistant to corrosion, durable, easy to clean, and suitable for its intended use.

- Unsuitable materials include:
  - Soft wood, which is absorbent, can splinter, and encourages cross-contamination and physical contamination
  - Copper, zinc, and aluminum, which may result in chemical hazards, especially when used with acidic food

Food equipment that cannot be dismantled, moved, or easily cleaned is also hazardous.

- Food equipment needs to be arranged and positioned carefully so that it is easy to access other equipment and nearby areas for cleaning and maintenance.

- The equipment's placement may need to allow for rapid dismantling and reassembly. The distance a piece of equipment is placed from a wall is determined by its size and ease of access for cleaning and maintenance. Facility managers should follow the manufacturer's specifications for equipment placement.

- Stationary equipment should be mounted on legs that elevate the equipment at least six inches off the floor.

- Stationary tabletop equipment should be on legs that elevate the equipment a minimum of four inches from the tabletop.

- Cracks or seams wider than $\frac{1}{32}$ inch where equipment is attached to a floor, wall, or tabletop must be sealed with a nontoxic, food-grade material.

- Sometimes equipment is designed so that it can be attached to a wall or other surface with a bracket (also known as cantilever mounted). This makes cleaning behind and underneath mounted equipment much easier.

Utensils must be corrosion resistant, smooth, and easy to clean.
Evgeny Litvinov, 2014/Shutterstock

UTENSILS

Utensils that contact food should be safe, corrosion resistant, waterproof, smooth, easy to clean, and sturdy enough to withstand frequent washing. Utensils should be resistant to damage such as chipping, scratching, and deformation. Wood utensils must be made of a close-grained hardwood, such as maple. Hardwood is an appropriate material for utensils such as cutting boards, rolling pins, salad bowls, and chopsticks.

Rubber and rubber-like materials are acceptable for use as utensils if they are durable enough to resist breaking, chipping, and scratching. Rubber utensils must withstand being cleaned and sanitized in a commercial dishwasher.

While cast iron is not an acceptable material for utensils, it may be used as a food contact surface for cooking. Cast iron may be used to serve food if the utensil is a part of the uninterrupted process that continues from cooking through service.

Once tableware and utensils are cleaned and sanitized, they must be stored properly, away from potential contamination. Soiled tableware and utensils should be kept separate from clean items. Clean items should be kept in a dry, stainless-steel storage area that is at least six inches (15.24 cm) above the floor. They should not be exposed to food or dust. When presented to customers, tableware should be prewrapped or stored, and only handles should be touched.

NON–FOOD CONTACT EQUIPMENT

Parts of some equipment—such as legs, housings, and supports—may not have any actual physical contact with food, but still need to follow basic guidelines in regard to sanitary design. The equipment may be splashed or spilled on or soiled by food. Non–food contact parts and equipment must be smooth, waterproof, corrosion resistant, easy to clean, and simply designed without ledges or hard-to-reach areas. Non–food contact surfaces should be kept clean and free from debris.

Store food equipment properly.
Svetlana Yudina, 2014/Shutterstock

PURCHASING EQUIPMENT

When choosing equipment for the facility, managers can reference the list of American National Standards Institute (ANSI), equipment compiled by organizations such as NSF International and Underwriters Laboratories (UL). These organizations develop standards for sanitary equipment design and environmental and public health. Approved equipment will be marked with the NSF or UL logos. Look for this mark on food service equipment before purchasing it. When choosing equipment, be sure to use only commercial-grade food service equipment. Common household equipment and appliances are not built for the heavy use that occurs in a food facility.

When considering the equipment needs of the food facility, there are several things food managers must keep in mind in addition to making sure the equipment is commercial grade:

- For large equipment, such as sinks, dishwashing machines, refrigerators, and walk-in coolers or freezers, there are many sizes and styles to choose from. Facility managers should consider how much space there is, how frequently the equipment will be used, how easy it will be to clean the equipment and the surrounding area, and how much routine maintenance will be required to ensure proper operation. It is also a good idea to check with local regulatory agencies for specific equipment requirements.
- Consider the location of plumbing and supply lines when deciding on the type of equipment to purchase and its placement.
- After purchase and installation of the equipment, always refer to the manufacturer's specifications for proper use, maintenance, and cleaning procedures.

MAINTENANCE GUIDELINES

It is not enough that food equipment is well designed and constructed. To avoid contamination, food equipment also must be properly maintained and used correctly.

Immovable equipment
erwinova/Shutterstock

Here are some guidelines for ensuring the good condition and proper use of equipment:

- Use all equipment according to the manufacturer's instructions. Never overload refrigerators or freezers, always maintain the right temperature, and keep doors shut.

- Immediately report any damage that could cause equipment or food to become unsafe or contaminated. For example, report any noticeable cracks or deterioration on door seals of refrigerators or freezers.

- Ensure that all equipment is effectively cleaned on a regular basis using the proper cleaning methods. This will help to avoid contamination and the buildup of soil and dirt, which allow bacterial multiplication and attract pests.

- Never use the same equipment for handling raw and ready-to-eat foods without thoroughly cleaning and sanitizing in between use in order to avoid cross-contamination.

- Color coding is another method food managers can employ to keep equipment hazard-free beyond cleaning and sanitizing. Use a color-coding system to reduce the risk of cross-contamination by ensuring that the same equipment is not used for different food types. Color coding can be used for many types of equipment, such as cutting boards, knife handles, work surfaces, cloths, protective clothing, and packaging material.

- Color-coding systems may differ from one facility to another, so it is important that staff is knowledgeable about the color code.

Learning Objective: Explain the importance of cleaning and sanitizing the food service facility.

Cleaning and sanitizing can greatly reduce the presence of microorganisms in food facilities. Food managers must ensure that proper cleaning equipment and supplies are available so that staff can adequately clean and sanitize the facility and equipment.

CLEANING

Ensuring that the food facility is clean and safe is the food manger's responsibility. Thorough and regular cleaning helps ensure that the facility is:

- A safe and pleasant place to work and eat in
- Free of contaminants and foodborne illnesses

Additionally, cleaning helps maximize efficiency of and prevent damage to food equipment. The purpose of **cleaning** is to remove food residue, dirt, grease, and other types of soil from surfaces. Cleaning methods include:

- Physical cleaning—for example, scrubbing
- Thermal cleaning—for example, the use of hot water
- Chemical cleaning—for example, the use of a **detergent**
- A combination of all of these

Employees should understand how to properly use cleaning chemicals, and which equipment or task is appropriate for the respective cleaners. This will prevent physical and chemical contamination of equipment, surfaces, and food.

CLEANING AGENTS

Many factors can have an effect on how cleaning should be done and which **cleaning agent** is most effective for the task at hand. Some of these factors include:

- **The type of material.** Some cleaning agents can damage certain surfaces.
- **The hardness or softness of the water.** Each type of water can cause a cleaning agent to react differently. Hard water can cause a buildup of minerals that may need to be removed using a specific cleaning agent.
- **The amount and type of soil that needs to be cleaned.** Large amounts of caked-on soil will take more time and work to remove.
- **The temperature of water being used.** Some cleaning agents work better with water at a specific temperature.

With any cleaning agent, follow the EPA-registered label use instructions. Make sure to check whether the cleaning agent is appropriate for use on a food contact surface.

Cleaning agents can be less effective if used improperly, and mixing cleaning agents can cause toxic fumes. Many of these cleaning agents are considered hazardous when used improperly:

- **Detergents**—remove dirt and grease.
- **Degreasers**—help dissolve dirt and remove it. Also known as solvent cleaners.
- **Acid cleaners**—also known as delimers, used to remove tarnish, alkaline discoloration, and corrosion from metals. They can also be used to remove hard-water deposits from dishwashers and serving tables.

Acid cleaners are also known as delimers.

zimmytws/Shutterstock

- **Abrasive cleaners**—used to physically scrub dirt and soil from surfaces. Abrasive cleaners can damage surfaces and make them hard to clean in the future, so care should be taken when using this type of cleaner.

Cleaning tools can be vehicles for physical, chemical, and bacterial contamination. To prevent contamination from cleaning tools and equipment, use the following guidelines:

- All tools that clean food contact surfaces should be cleaned and sanitized after use.
- Where appropriate, tools should be left to air-dry.
- Staff should never leave cleaning materials in dirty buckets or soaking in water overnight.
- Just like cleaning chemicals, cleaning tools should be kept separate from food items.
- Designate tools for time/temperature control for safety (TCS) food areas, and separate TCS food preparation tools from raw food tools.
- Ready-to-eat food areas must be cleaned before raw food areas.
- All areas must be cleaned from top to bottom.
- Color coding of tools is recommended as an effective way to reduce the risk of cross-contamination. That way, for example, tools used to clean the restroom aren't used to clean the meat slicer.

Cleaning supplies, such as brooms, mops, and wiping cloths, should all be stored away from any food contact surfaces. Cleaning tools and supplies should be air-dried unless they are stored in a sanitizing solution. Cleaning tools that are used in non-food contact areas should not also be used in food-contact areas. Only designated, sanitized cleaning tools should be used in food areas.

Cleaning chemicals must never be stored together with food or come in contact with food.

SANITIZER

The purpose of **sanitization** is to reduce the number of microorganisms to a safe level. Sanitization is not the same as disinfection. **Disinfection** means the process of destroying microorganisms to a level in which all except for bacterial spores are killed; the chemical used is called a disinfectant. Due to the increased costs and hazards of disinfectants, food establishments use sanitizers for food contact areas. Sanitization methods include the use of hot water, steam, or a chemical sanitizer.

Sanitize refrigerator doors.
Rob Marmion/Shutterstock

In order for these processes to be effective, the equipment or surface must be thoroughly cleaned and rinsed before being sanitized. The sanitizers need to be in contact with the equipment or surface for a sufficient length of time, referred to as contact time. All equipment and surfaces need to be cleaned regularly, but only some need to be sanitized.

Sanitization is normally restricted to:

- Food contact surfaces, such as cutting boards and knives
- Hand contact surfaces, such as faucets or refrigerator door handles
- Cleaning materials and equipment

If these surfaces are not sanitized, then food handlers may be introducing food safety hazards into the operation. However, if equipment or surfaces are un-necessarily sanitized—for example, non–food contact surfaces, such as floors or ceilings—then food handlers will be wasting time and materials. All traces of cleaning agents should be removed before sanitizing a surface. Cleaning agents left on a surface could contaminate the food or cause the sanitizer not to work properly.

Food surfaces that come into contact with TCS food must be cleaned and sani-tized throughout the day, at least every four hours. Equipment, food contact surfaces, and utensils should also be sanitized:

- When changing to a different type of raw animal food, unless the following raw animal food requires a higher cooking temperature than the previous type
- When changing from working with raw foods to working with ready-to-eat foods
- When changing from raw fruits and vegetables to TCS foods
- Before using or storing a food temperature–measuring device
- At any time during the operation when contamination may have occurred

A procedure should be in place for employees that addresses responding to and minimizing the spread of contamination.

HEAT METHOD

Hot water is one of the most commonly used sanitizers. It can be used by cleaning manually or by machine.

When using hot water to manually sanitize objects, immerse the object for at least 30 seconds in water that is a minimum of 171°F (77°C). It may be necessary to add a heat-ing element to the sink to achieve the correct temperature.

High-temperature dishwashing machines also use hot water, at 180°F (82°C) to sanitize objects. Closely monitor the temperature of the water to confirm that the dishwashing machine is working properly. If the water is too hot, it could vaporize before it sanitizes the objects.

CHEMICAL METHOD

Sometimes using the heat method is not appropriate for sanitizing an object. Chemicals can also be used for sanitizing when cleaning manually or by dishwashing machine. Sanitizing chemicals must be approved for use on food contact surfaces.

The chemical sanitizers most commonly used in the food community are:

- Iodine
- **Quaternary ammonium compounds**, also known as quats
- Chlorine

30 SEC

171°

To use the heat method of sanitization, an object must be immersed in water that is at least 171°F (77°C) for 30 seconds.

In order to work properly, a sanitizing solution using iodine must also have:

- A minimum temperature of 68°F (20°C)
- A pH of 5 or less, or a pH no higher than the level for which the manufacturer specifies the solution is effective
- A concentration between 12.5 mg/L and 25 mg/L

A quaternary ammonium compound solution must:

- Have a minimum temperature of 75°F (24°C)
- Have a concentration as specified under 21 CFR 178.1010 and as indicated by the EPA-registered label use instructions included in the labeling
- Be used only in water with 500 mg/L hardness or less or in water having a hardness no greater than specified by the EPA-registered label use instructions

A chlorine solution must have a minimum temperature based on the concentration and pH of the solution as listed in the following chart:

Concentration Range	Minimum Temperature	
mg/L	pH 10 or less °F (°C)	pH 8 or less °F (°C)
25–49	120° (49°)	120° (49°)
50–99	100° (38°)	75° (24°)
100	55° (13°)	55° (13°)

For chemical sanitizers to work effectively, several guidelines must be followed:

- Cleaning chemicals must always be used according to the EPA-registered label use instructions and company procedures.
- They must always be used at the recommended concentrations and must never be mixed or diluted.
- A sanitizer test kit should be used to check the concentration of the sanitizer; each type of sanitizer requires a different type of test kit.
- Sanitizers must be used at the proper temperature and be in contact with the surface for the required amount of time.

DESIGNING A CLEANING PROGRAM

Creating a cleaning schedule is the best way to ensure that a food facility is properly cleaned. All employees must know and understand their responsibilities for cleaning, and managers must ensure that cleaning is effectively carried out in all areas of the food business. The easiest way to do this is to create a written cleaning schedule, or Sanitation Standard Operating Procedure (SSOP).

Cleaning schedules must clearly specify details regarding the surface or equipment to clean. It is not enough that cleaning be carried out as specified in the schedules. In addition to following a cleaning schedule, employees should adopt a clean-as-you-go policy to prevent dirt from accumulating. A clean-as-you-go policy (for example, by mopping up spills immediately) will contribute to a safer work environment and prevent slips, trips, and falls.

Daily General Cleaning Schedule				Date:	05/06/16
Area to clean	**How to clean**	**Cleaning supplies**	**Times**	**Staff Initials**	**Mgt. Initials**
Floors (daily and as needed)	Sweep, mop	Approved sanitizer	2	M.A/C.K	L.B.
Dry Storage (daily and as needed)	Sweep, mop	Approved sanitizer	1	C.K.	L.B./F.D.
Prep Areas (daily and as needed)	Wash, rinse, sanitize	Warm soapy water and 200 ppm sanitizer	3	M.A/C.K /S.E.	L.B.
Hood Grease Pan (daily and as needed)	Clean with degreaser, wash with dishwasher	Warm soapy water and 200 ppm sanitizer	2	S.E./C.K.	L.B.
Hood Filter (daily)	Soak in degreaser, rinse, air dry	Degreaser	1	C.K.	F.D.
Storage Bins (daily and as needed)	Use a clean, damp cloth to wipe exterior	Warm soapy water and 200 ppm sanitizer	1	S.E.	L.B.
Trash Bins (daily and as needed each shift)	Use a clean, damp cloth to wipe exterior and interior	Warm soapy water and 200 ppm sanitizer	3	M.A/C.K /S.E.	L.B./F.D.
Walk-in Cooler (daily and as needed)	Sweep, moop; wipe outside and inside	Approved sanitizer`	2	M.A/C.K	L.B.

Written cleaning schedules help employees know their responsibilities.

SANITATION STANDARD OPERATING PROCEDURES (SSOPs)

For each surface and piece of equipment, food managers need to provide standard operating procedures that specify:

- What to clean
- How to clean and sanitize it
- How much time the task will require
- What standard is required
- What chemicals to use and the amounts needed
- What contact time is required
- What equipment should be used, including references to color coding
- Where and how to store chemicals and equipment after the procedures are completed

Written cleaning and sanitizing procedures must specify the safety precautions employees should follow. This could include wearing appropriate protective clothing or taking measures to protect food, such as covering open food and other surfaces. It should be clear who is responsible for monitoring and recording that equipment is clean and safe to use.

It is the facility manager's responsibility to develop cleaning schedules or to delegate this task to an appropriately trained staff member. Employees must become familiar with the facility's cleaning schedules and procedures and know what is required of them.

LESSON 4 | WASHING FACILITIES

Learning Objective: Describe the various washing facilities found in a food establishment.

HAND-WASHING STATIONS

Hand-washing stations are required in restrooms, dishwashing areas, and food preparation and service areas. All stations must be fully functional and have hot and cold water, soap, a drying method, and trash receptacles. A sign should be posted directing employees to wash their hands before returning to work. Staff facilities may include lockers, restrooms, and break rooms. First aid equipment must be kept away from food, food storage, and food production.

The design of the restrooms is especially important in food facilities. Of particular importance is the location of sinks for hand washing. The ideal place is between the toilet and the door. Restroom doors must not open directly into food preparation areas. It is vital that both staff and public restrooms are kept clean to avoid contamination and the spread of bacteria in the facility.

A pedal-operated hand sink enables hands-free operation. A splash guard divider prevents contamination between the hand sink and the adjoining warewashing sink area.

DISHWASHERS

Warewashing is simply the washing of the wares, or in the food industry, food contact materials. Warewashing can be done either mechanically or manually. Dishwashers can help save time by cleaning and sanitizing many items at once. Always follow EPA-registered label use instructions when operating a machine, and keep it in good repair.

Follow the instructions below when using any washing facilities:

- Monitor and maintain the correct water temperature for washing and sanitization.
- Rinse food and soil from utensils and objects before placing them into the dishwashing machine.
- Arrange all items so that water contacts the entire surface.
- Check that all food and soil is removed from items when the machine is finished.
- Clean the machine at least once every 24 hours, and monitor or clean it throughout the day as needed.
- Items that cannot be washed or sanitized in a dishwashing machine must be washed manually.

Commercial dishwashing machines are available in two types:

High-temperature machines

High-temperature machines use hot water and detergent to clean and sanitize. The temperature of the wash cycle in a high-temperature machine can vary according to the type of machine. Checking the EPA-registered label use instructions and the FDA Food Code can help determine the proper wash temperature for a machine. The final rinse temperature should be 180°F (82°C) to ensure sanitization. If the final rinse temperature is over 195°F (91°C), then it is too hot and the water will vaporize before the items are sanitized.

Chemical sanitizing, or low-temperature, machines

Chemical machines operate at a lower temperature than the high-temperature machines and use a chemical sanitizer. The temperature that the machine operates at will be determined by the type and concentration of chemical sanitizer uses, but it should not be lower than 120°F (49°C) for the wash cycle.

THE FOOD EQUIPMENT SINK

A three-compartment sink is the most common type of sink used. Two-compartment and four-compartment sinks are sometimes used as well. Check local regulations for specific information on sink compartment guidelines. The sink compartments should be large enough that any item can be fully immersed if it needs to be washed by hand.

DISHWASHING STATION SETUP

A manual dishwashing area should be set up with specific arrangements in mind:

- An area should be designated for storing dirty and soiled items, and also a separate area to air-dry clean and sanitized items.
- There should be a place to remove food from items before they are washed.
- The station should include a hand sink, so that hands can be washed between handling dirty and clean dishes.
- The station should include a thermometer to measure water temperature and a clock with a second hand to monitor sanitization times.
- If using a chemical sanitizer, test kits must be available to check the concentration of the sanitizing solution.

Keep the dishwashing area organized.
Kondor83, 2014/Shutterstock

PROPER DISHWASHING PROCEDURE

When washing items manually:

1. Start with a clean and sanitized sink and dishwashing station. Remove as much of the visible soil and food as possible before washing in the three-compartment sink.

2. Use the first compartment to wash items with a detergent. Make sure that the water temperature is at least 110°F (43°C). Replace the detergent and the water in this compartment as needed.

3. The second compartment is used to rinse the detergent off of the item and prepare for sanitization. Spray-clean the objects or immerse them in water to remove the detergent. The water in this compartment should be kept clean and replaced when needed.

4. The third compartment holds a sanitizing solution. It may need a device to keep the temperature of the sanitizing solution at a specific temperature. The temperature of the solution and the contact time for the solution will be determined by the type and concentration of chemical sanitizer used.

5. For hot-water sanitization, immerse the object for at least 30 seconds in water that is at least 171°F (77°C).

6. Once an item is sanitized, it should be allowed to air-dry.

CLEANING IN PLACE

Food facilities should be designed and constructed to promote and enable effective cleaning and sanitizing. Sometimes this is not a straightforward process.

CIP—or cleaning in place—is necessary when equipment cannot be dismantled or moved. This process is most common in dairy premises, breweries, and beverage manufacturing, or when the equipment is large, floor mounted, or sealed. CIP equipment should be self-draining or capable of being completely drained of cleaning and sanitizing solutions.

Equipment that cannot be disassembled should have access points that allow inspection of interior food contact surfaces to make sure it is properly sanitized. Prior to cleaning CIP equipment, unplug electrical equipment and remove detachable parts for manual wash.

A typical CIP sequence consists of five steps:

1. A prerinse, to remove soil in the pipes
2. Detergent circulation, to remove residual debris and dissolve grease
3. An intermediate rinse with water
4. Sanitization, to destroy the remaining organisms to a safe level
5. Air-drying

Reassemble any parts removed for manual washing and re-sanitize the parts that are handled during reassembly. Follow the EPA-registered label use instructions and the FDA Food Code for clean-in-place equipment instructions. Take extra care with refrigerators and dangerous equipment such as meat slicers.

PLUMBING

Learning Objective: Explain the importance of safe drinking water in a food service facility.

WATER

All food premises must have a satisfactory, constant supply of drinking water. Only drinking water, also known as **potable water**, can be used in food preparation and for cleaning food or food contact areas. This includes water used to clean equipment and in prep sinks, dishwashers, and steam tables. Sources of drinking water include public and private water systems that are properly maintained and tested or sampled annually. Types of water that could be used in the facility include bottled drinking water, properly maintained water pumps, hoses, pipes, water transport vehicles, and water storage containers.

NONDRINKING WATER

The only exceptions to using drinking water are if the water is intended for use in air-conditioning, fire protection, or non-food equipment cooling. Nondrinking water must not come into contact with any food or food contact surfaces. Nondrinking water pipes must be labeled as such.

PRIVATE WATER SUPPLY SYSTEMS

If the food facility uses a private water supply system, it should be inspected at least annually. Check with local regulatory agencies for specific information on testing and inspections. A copy of the most recent sample report should be kept and filed.

If the construction of the food premises is new, the water system should be flushed and sanitized before becoming operational to remove any possible contaminants. There are other times when flushing the system will be necessary—the system should be flushed after a repair or modification to the system has been made, and after an emergency situation has occurred, such as a flood.

EMERGENCY GUIDELINES

Occasionally an emergency situation occurs, such as a hurricane or flood, that affects the drinking water supply. If there is a need to maintain service, follow these safety guidelines:

- Listen for announcements from local authorities that advise whether the water is safe to use.

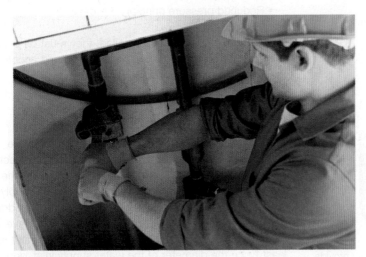

The facility needs to follow plumbing codes.
Phovoir/Shutterstock

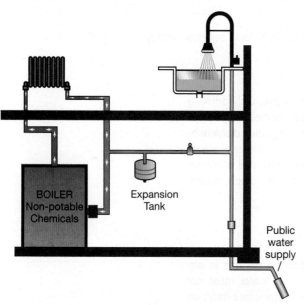

Back pressure can take place when a piece of equipment generates pressure greater than the drinking water supply.

- If the water is deemed unsafe, boil it to kill harmful bacteria and parasites that may be present. Most organisms are killed in water that is held at a rolling boil for 60 seconds.

- It is possible to treat water with bleach, chlorine tablets, or iodine tablets. If any of these water treatment methods are attempted, check the label for directions on how to properly use the products. Be aware that many parasitic organisms will not be killed using these methods. Boiling is the best method to use.

- Boiling will not remove chemical contaminants. If it is not clear whether or not the water has been chemically contaminated, use bottled water from a safe source.

FOOD FACILITY PLUMBING

An important function of the plumbing system in food facilities is to prevent drinking water from mixing with nondrinking water. When the two types of water come into contact with each other, a **cross-connection** has occurred. Cross-connections are known to cause foodborne illness outbreaks.

A cross-connection is defined as an actual or potential connection between a drinking water supply and a nondrinking source. In a cross-connection, it is possible for a contaminant to enter the drinking water supply. This reverse flow is called **backflow**, which is the flow of water or other liquids, mixtures, or substances into a drinking water system from any source other than the intended source.

Backflow occurs when wastewater develops a reverse flow and enters a drinking water source within the same system. Backflow is usually caused by a pressure drop in the facility's water supply that creates a negative supply pressure; this specific form of backflow is called **back siphonage**. Backflow due to **back pressure** can take place when a device such as a boiler generates pressure greater than that of the drinking water supply. In the absence of an air gap or an approved reduced pressure device, the boiler can build up pressure and force water from the boiler into the drinking water supply.

The plumbing code requires that systems be installed using appropriate backflow prevention. There are several different types of approved backflow prevention devices available:

- Barometric loops
- Vacuum breakers (atmospheric and pressure types)
- Dual-check valves with intermediate atmospheric vents
- Dual-check valve assemblies
- Reduced pressure principle devices

If any of these devices are installed, they must be placed where they can be properly serviced and maintained according to the manufacturer's specifications.

AIR GAPS

Although backflow prevention devices are available, the most simple way to prevent backflow from a cross-connection is the non-mechanical method: the **air gap**. An air gap is the air space that separates the opening to a drinking water supply line from a potentially contaminated source, such as a floor drain. An air gap between the water supply inlet and the flood level rim of the plumbing fixture, equipment, or non-food equipment must be at least twice the diameter of the water supply inlet, and it may not be less than 1 inch (25 mm).

Air Gap

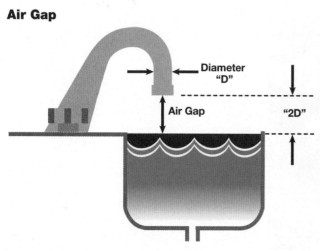

Devices such as air gaps are used in plumbing systems to provide backflow prevention.

LEAKS

Another part of a facility's plumbing system is overhead pipes that may carry wastewater. These pipes should be clearly marked as nondrinking water. If these pipes leak, contamination of food and food preparation surfaces will occur. Even if the overhead pipes contain drinking water, a leak is still a problem. Condensation that forms on the leaky pipe can drip down onto food and food preparation surfaces. All of the lines should be checked and serviced immediately if a leak occurs.

GREASE TRAPS

Grease traps (also known as grease interceptors, grease recovery devices, and grease convertors) are designed to separate most greases and solids in wastewater before they flow into the sewage system. As the facility wastewater flows into the grease trap and cools, the grease, fats, and oils harden and float to the top of the trap. The remaining water continues to flow into the public sewer, leaving most of the solids behind.

SEWAGE AND WASTEWATER

Food facilities should have an efficient drainage system to remove sewage and wastewater quickly without flooding. Sewage and wastewater are highly contaminated, and contact with food or food preparation surfaces must not occur. The drainage system should allow sufficient access for cleaning in the event of blockages and be constructed to prevent pests from trying to live in the system. If grease traps are used in the facility, they must allow for easy access, and they must be maintained and serviced by a licensed plumber. The drainage system must empty directly into the public sewage treatment facility or another approved disposal system that operates according to legal specifications.

KEY TERMS

ADA   The Americans with Disabilities Act is a federal civil rights law that prohibits discrimination against people with disabilities.

Air gap   The vertical air space that separates the end of a supply line and the flood level rim of a sink, drain, or tub. It is one of the cheapest and most reliable methods of backflow prevention.

Back pressure   Backflow due to a device, such as a boiler, that generates pressure greater than that of the drinking water supply.

Back siphonage   Backflow due to a sudden drop in the supply pressure of a water main, causing contaminated water to reverse-flow back into the water main.

Backflow   The reverse flow of water from a contaminated source to the drinking water supply. It can occur from back pressure or back siphonage.

CIP   Stands for cleaning in place. This cleaning process is necessary when equipment cannot be dismantled or moved. It involves the circulating of non-foaming detergents and disinfectants, or sanitizers, through assembled equipment and pipes, using heat and mostly turbulence to attain a satisfactory result.

Cleaning   The process of removing soil, food residues, dirt, grease, and other objectionable matter; the chemical used to do this is called a detergent.

Cleaning agent   A chemical compound, such as soap, that is used to remove dirt, food, stains, or other deposits from surfaces.

Cross-connection   The mixing of drinking and contaminated water in plumbing lines.

Detergent   A chemical or mixture of chemicals made of soap or synthetic substitutes. It facilitates the removal of grease and food particles from dishes and utensils and promotes cleanliness so that all surfaces are readily accessible to the action of sanitizers.

Disinfection   The process of destroying microorganisms to a level in which all except for bacterial spores are killed; the chemical used is called a disinfectant.

Potable water   Water that is safe to drink; an approved water supply.

Quaternary ammonium compounds   A common example of chemical disinfectants; also referred to as quats.

Sanitization   Use of chemicals or heat to reduce the number of microorganisms to a safe level.

Warewashing   The washing of the wares, or in the food industry, food contact materials.

ASSESSMENT QUESTIONS

1. When surfaces are in contact with time/temperature control for safety foods, what is the maximum amount of time allowed between sanitizations?
   a. Twelve hours
   b. Eight hours
   c. Four hours
   d. One hour

2. For manual hot water sanitization, items should be immersed in water for 30 seconds at what temperature?
   a. 185°F (85°C)
   b. 171°F (77°C)
   c. 121°F (50°C)
   d. 98.6°F (37°C)

3. What is the easiest way to prevent backflow from a cross-connection?
   a. Barometric loops
   b. Air gaps
   c. Reduced pressure principle devices
   d. Double-check valve assemblies

4. Hand-washing stations are NOT required in:
   a. Food service areas
   b. Break rooms
   c. Dishwashing areas
   d. Food preparation areas

5. Where is hard-surface flooring suitable for use?
   a. Dish area
   b. Walk-in refrigerator
   c. Food preparation area
   d. Restroom

6. What is the minimum lighting requirement for a food preparation area?
   a. 108 lux (10 foot-candles)
   b. 215 lux (20 foot-candles)
   c. 540 lux (50 foot-candles)
   d. 620 lux (60 foot-candles)

7. Stationary kitchen equipment must be how far off of the floor?
   a. Two inches
   b. Five inches
   c. Six inches
   d. Eight inches

8. When dirty water reverses into the clean water supply, that is called:
   a. Backflow
   b. Cross-connection
   c. Non-drinking water
   d. Air gap

9. Which of these is an acceptable way to use nondrinking water?
   a. Cleaning equipment and utensils
   b. Mixing drinks
   c. Cleaning non–food contact surfaces
   d. Air-conditioning

10. What occurs when drinking water mixes with nondrinking water?
    a. Cross-contamination
    b. Cross-connection
    c. Backflow
    d. Back siphonage

QUESTIONS FOR DISCUSSION

1. Imagine that you are working with a team of contractors to build a brand-new restaurant, and you are in charge of the layout. What are some ways that you can promote food safety in your establishment through the facility layout?

2. What are some ways to prevent contamination from food equipment in your establishment?

3. Imagine that you are the manager at a very busy Italian restaurant. How can you and your employees prevent contamination during cleaning or by cleaning tools?

4. What are some of the steps in the proper dishwashing procedure for washing items manually?

5. Discuss the differences between the terms *cross-connection*, *backflow*, and *back siphonage*. How are these terms similar?

PURCHASING AND STORING FOOD

The path that food follows through a food facility is known as the **flow of food**. Typically, the flow of food through a retail establishment begins with purchasing the food and continues with storing, preparing, holding, serving, and selling the food to the consumer.

The flow of food involves many steps, called operational steps, which can include:

- Purchasing
- Receiving
- Storing
- Preparing
- Cooking
- Cooling

- Reheating
- Holding
- Assembling
- Packaging
- Serving
- Selling

Although the terms used to refer to the various operational steps may vary among different types of retail food establishments, identifying the operational steps that food follows as it moves through a facility can help managers understand where cross-contamination can occur.

This chapter focuses on the early operational steps that take place before food preparation even begins.

After reading this chapter, you should be able to:

- Evaluate food safety controls of the supplier.
- Identify the potential hazards associated with food delivery.

- Describe the actions needed to avoid hazards in food storage.

Be sure to use a reputable supplier.
Alistair Berg/Getty Images

LESSON 1 | PURCHASING

Learning Objective: Evaluate food safety controls of the supplier.

Purchasing is the first step in the flow of food. Food must be purchased from an approved, reputable supplier. A reputable supplier is one with an outstanding record in food safety.

Food products should be purchased from approved sources that meet all local, state, and federal regulations, and operate under government inspection. This is especially important with produce, dairy, egg, and seafood suppliers. Sealed foods must be purchased from a processing plant governed by a regulatory agency, such as the FDA or the USDA.

A reputable supplier can be identified in many ways. Some things suppliers should be able to provide or demonstrate include:

- A **food safety management system**, or a HACCP program
- Procedures to ensure food safety
- Food safety training for employees
- Good condition of delivered food
- High standards of the driver and vehicle
- No food prepared in a private residence
- Impeccable premises, with full records of products going in and out
- A hygienic delivery vehicle
- Refrigerated vehicle when transporting chilled or frozen goods
- No materials in the vehicle that could contaminate the food (such as toxic chemicals/equipment)

The delivery driver should have impeccable personal hygiene and be trained on how to deliver goods correctly. Food delivery drivers must know how to avoid cross-contamination of foods within a large, mixed delivery.

The food manager or his or her person in charge should regularly monitor deliveries to ensure quality and food safety. The supplier must immediately address problems.

Checking a supplier's HACCP documentation and comparing it to the facility's process can help identify potential issues and recognize the supplier's food quality and food safety practices. Research the suppliers, and ask other people in the field about potential or current suppliers. If a supplier has previously delivered spoiled or damaged food, explore new vendors. Whenever possible, inspect a prospective supplier's warehouse or plant to ensure that it meets high standards for food safety.

Some food items, such as frozen raw chicken or raw shell eggs, may present an unacceptable risk. Choosing alternative ingredients, such as fresh cooked chicken or pasteurized eggs, can reduce this risk. **Pasteurization** is the process of applying heat to destroy pathogens. Food items without appropriate packaging are at a greater risk of contamination. Fish and shellfish are required to be caught and harvested legally and commercially. Shellfish is required to have certification because of the possibility of environmental contamination. Shellfish must arrive with identification tags, and the tags must be kept on file for 90 days. These tags make it possible to trace any shellfish responsible for a foodborne illness back to the initial date and location of sale. For this reason, it is vital that batches of shellfish do not mix.

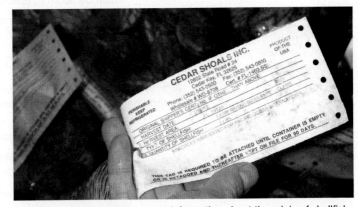

Shellstock tags provide important information about the origin of shellfish.

SUPPLIERS: TRANSPORTATION AND DELIVERY

Learning Objective: Describe proper procedures for the transportation and delivery of food.

There are two significant safety hazards food managers will potentially face when dealing with the delivery of food:

- Contamination of food by physical contaminants, chemicals, or bacteria. This can occur either before or at the point of delivery.
- The multiplication of bacteria within the food due to improper temperatures, either en route to the establishment or at the point of delivery. Delivered food items should never be left outside for prolonged periods prior to storage.

DELIVERIES

By specifying to the supplier exactly what the facility's expectations are regarding delivery times and the condition of food items upon receipt, food managers can reduce the risk of receiving unsuitable food. Request that deliveries be made when staff is free to inspect them. Employees must verify that deliveries of food during non-operating hours are from approved sources. Deliveries must be placed into appropriate storage locations, maintained at the required temperature, protected from contamination, and kept safe for consumption.

Having a separate unpacking area can help ensure that contaminants on the surface of a food item can be safely spotted before moving it into a food preparation area. Separate unloading/unpacking areas can help prevent any potential contamination from entering the facility during a delivery.

Each of the following must be observed carefully and may raise a flag with food managers to investigate the supplier:

- Boxes and packaging must be clean and free from any signs of pest infestation, contamination, dampness, or damage. If any of these are spotted, the product needs to be rejected.
- Canned and vacuum-packed foods must be checked closely because *Clostridium botulinum* can grow in foods packaged this way. Reject food if packages are swollen or show signs of leaking. Reject canned food that is rusted or dented, or if one end of a can pops when the other end is pushed.
- Check delivered food items to ensure they are not out of date. Additionally, care must be taken to ensure items will not expire before their intended use dates.
- Any products that are not labeled must be rejected.

TEMPERATURE CONTROL

Proper temperature control must be maintained, even during delivery. Some things that the food handlers should ensure when receiving shipments of delivered food include:

- For chilled or frozen goods, an appropriate refrigerated vehicle is required.
- Raw eggs must be delivered via equipment that maintains an **ambient temperature** at or below 45°F (7°C).
- Refrigerated foods must be received at 41°F (5°C) or colder.
- Refrigerated meats, poultry, and seafood may be received at temperatures as low as 28°F (−2°C).

- Fresh fish should optimally arrive alive, be packed on ice, and have an internal temperature between 32°F and 41°F (0°C and 5°C). Check local regulations for specified receiving temperatures other than 41°F (5°C).
- Exceptions to required delivery temperatures may apply for live molluscan shellfish and/or shell eggs.
- Frozen food should be received below 0°F (−18°C).
- Reject open containers and any foods that show signs of thawing and refreezing.
- Clumping, ice crystals, and puddles from frozen water at the bottom of a container can indicate thawing and refreezing.
- Deliveries of chilled food above 41°F (5°C) and frozen food that is not still frozen must be rejected.

Whoever checks the delivery should ensure that any shortcomings are communicated to the delivery driver and stated on the delivery note. The individual assigned to check in food must inform management immediately if there are issues, as the incident will require further investigation of the supplier. It may even result in a change of supplier.

Whoever checks a delivery needs to check the delivery note or purchase order to ensure that the quantities delivered match the note or purchase order. The same checks are needed for the type and quality standards of the food delivered. Reject food that does not meet the standards of the establishment. When rejecting food, make sure that staff keeps the rejected product away from other food, records the rejection on the delivery receipt, informs the delivery person of the rejected product, and gets a signed adjustment or credit slip from the delivery person prior to removing the product. Any shortcomings should also be reported to management and investigated. Check delivery-monitoring records regularly.

After delivery, all food items must be thoroughly checked; outer packaging must be removed and food stored correctly to prevent bacterial multiplication. Time/temperature control for safety (TCS) foods should be stored quickly, within 15 minutes of delivery.

Check refrigerated delivery truck temperatures.
Robert Pernell, 2014/Shutterstock

Learning Objective: Describe the actions needed to avoid potential hazards in food storage.

The three main types of delivered goods received by food establishments are: refrigerated items, frozen items, and dry goods. The food manager must strictly follow the proper guidelines for storing these items to prevent spoilage and contamination. Refrigerated items should be stored first, followed by frozen foods, and then dry goods.

Newly delivered food items should be stored below or behind older stock. Stored goods must be correctly rotated. Correct **stock rotation** helps to ensure that the food used is safe and of a good, consistent quality. Stock rotation can help to avoid spoilage, reduce the risk of pest infestation, and minimize the risk of food going out of date. Remember the rule—"first in, first out," or **FIFO**. If employees follow this rule, then older food will always be used first and the establishment will minimize waste and save money.

Written stock control records are also recommended. Written records help maintain correct stock levels and determine when out-of-date food should be discarded. It helps to identify necessary corrective actions, such as further training for staff on date labels or stock rotation principles. Check the date and appearance of all TCS foods and perishable foods daily. **Perishable foods** are those foods subject to spoilage or decay.

Out-of-date foods must be removed and destroyed, or stored separately pending appropriate action. Require employees to notify management in any scenario that has expired foods.

REFRIGERATED STORAGE

The major hazards of refrigerated food storage are contamination of foods by raw foods, and the multiplication of bacteria or spoilage organisms if temperatures are too high or storage is prolonged. There are many controls that can be put into place to avoid the hazards associated with refrigerated storage.

Follow controls closely to avoid hazards with refrigerated foods.
Noah Strycker, 2014/Shutterstock

The following are examples of controls that can minimize hazards for refrigerated foods:

- Nearly all TCS foods must be stored at or below 41°F (5°C). Some foods, such as fresh fish, should be stored at lower temperatures. The maximum shelf life for any TCS food and ready-to-eat food prepared in the establishment is seven days. Any TCS foods or ready-to-eat foods stored longer than seven days or at an incorrect temperature must be discarded. In order to maintain this operating temperature, refrigerators must be located in well-ventilated areas away from heat sources.

- Refrigerators cannot be overloaded, and should have open shelving.

- Congested shelving prevents cold air from circulating, which can raise the temperature of the refrigerator and food stored in it. Do not use refrigerators for cooling food.

- Alarmed units are recommended, to automatically inform of high humidity values and unacceptable temperatures. High humidity encourages the growth of spoilage bacteria. Sophisticated automatic monitoring, data loggers, and alarm systems are becoming more common.

- Keep refrigerator doors closed as much as possible to help maintain proper temperatures. Install cold curtains on frequently used walk-in refrigerators.

- Ideally, ready-to-eat foods and raw foods should be kept in separate refrigerators. If in the same unit, raw food such as eggs, meat, and poultry must always be placed below TCS food and ready-to-eat food, such as lettuce and tomatoes, to avoid cross-contamination.

- Food storage order from top to bottom is: ready-to-eat foods, fruits and vegetables, fish and seafood, beef and pork, ground meats, and then poultry.

- Stock should be rotated so that older food is always in the front of the refrigerator and used first.

- Refrigerators must be cleaned and sanitized regularly, to prevent the accumulation of dirt or physical contaminants. Care must be taken to prevent chemical contamination of food items or refrigerator surfaces while cleaning and sanitizing.

Rotate stock in refrigerated storage.
Jonathan Feinstein, 2014/Shutterstock

When organizing a refrigerator, food should be stored in the following order to prevent cross-contamination (from top to bottom): ready-to-eat foods and fully cooked foods; whole raw seafood, fish, and eggs; whole raw beef and whole raw pork; raw ground meat and fish; raw poultry (chicken, turkey, duck).

- Properly cover food to prevent drying out, cross-contamination, and absorption of odor. For example, storing uncovered onion slices in a refrigerator will cause other foods to take on the onion odor. Covering an open can and placing it in the refrigerator is not acceptable. Empty unused contents of a can into a suitable covered storage container and apply the proper labeling.

- Fit all refrigerators with accurate thermometers that use liquid crystal displays. Displayed temperatures and temperature of the food in the warmest part of the refrigerator (which is usually the top) may not always match. Be sure to check these areas.

- Check the display temperature every time the refrigerator is used. Check the temperature of food, or a food substitute, with an accurate, calibrated digital thermometer at the start of the day, at the end of the day, and whenever the display reading is unacceptable.

- Maintain a record of food temperatures and check it weekly.

Food safety managers should audit refrigerators regularly. Audits address excessive temperatures and humidity, as well as how staff loads and cleans the refrigerator. Audits should check for any electrical problems, refrigerant leaks, or damage to door seals, as they can contribute to an increase in the refrigerator temperature. A maintenance contract is recommended to ensure that refrigerators are kept in good repair.

Unfit or contaminated food should be discarded immediately. All food in refrigerators should be in appropriate covered containers. Any open cans or damaged containers should be removed from the refrigerator and rejected. Food must never be stored in empty chemical containers. The base of storage containers cannot come into contact with open food. Out-of-date food must be rejected at once. Check all "use-by" date labels daily, and rotate stock appropriately. The **use-by date** is the last date recommended for the use of the product if it is to be used at peak quality. The manufacturer of the product usually determines the date. A retail facility may not sell any food past the use-by date.

FROZEN STORAGE

The major hazard associated with frozen food storage is the multiplication of dormant bacteria that takes place when the temperature rises above 14°F (–10°C). Unfortunately, freezing does not kill bacteria. Frozen food should never be left out at room temperature. After frozen food is delivered and inspected, it should be immediately placed in the freezer. Freezer temperature should be maintained at 0°F (–18°C). Do not refreeze food.

In order to maintain the correct operating temperature:

- Freezers must be located in well-ventilated areas away from heat sources.
- Fit freezers with accurate thermometers, and check temperatures daily.
- Staff should record and review temperature logs regularly.
- Do not store food above the freezer load line.
- Keep freezers closed as much as possible to help maintain proper temperatures.
- Install cold curtains on freezers that are frequently used.
- Do not store warm food, as this can result in other foods thawing.

Label food before freezing the package with the content and its use-by date, and observe the principle of proper stock rotation.

Raw meat, poultry, and seafood can be stored with or above ready-to-eat food in a freezer if all of the items have been commercially processed and packaged.

It is essential to follow the manufacturer's instructions on storing foods, as food quality gradually deteriorates, even in the best of freezers. For example, vegetables, fruit, and most meat can be frozen for up to 12 months, whereas pork, sausages, fatty fish, butter, or soft cheese should only be stored for up to six months.

Regular audits of freezers address temperatures, food load, and cleaning. Audits should also involve checks for any electrical problems, refrigerant leaks, and damage to door seals. A maintenance contract is recommended to ensure freezers are kept in good repair.

Freezers are most efficient if they are frost-free. Before manually defrosting a freezer, move the contents to another freezer. All "best before" date labels should be checked weekly, and stock rotated appropriately. Any out-of-date food should be discarded.

DRY STORAGE

There are two main hazards associated with dry food storage. The first hazard is physical contamination, which occurs from objects brought in on delivery packaging, such as staples or cardboard. The second hazard is chemical contamination, which comes from rusty cans or chemicals kept in dry storage. Multiplication of food-poisoning bacteria or food spoilage microorganisms can become a hazard if the food becomes damp. Do not store food in hallways, utility rooms, restrooms, or sleeping and dressing rooms.

Dry storage areas need to be kept clean and well maintained. Managers should take steps to ensure safe dry food storage areas, including the following:

Dry storage areas should be kept clean and well-maintained.

- Maintain the dry storage temperature between 50°F and 70°F (10°C and 21°C).
- Keep area well lit and ventilated.
- Use easy-to-clean shelving.
- Use slotted shelving units to improve air circulation.
- Store food at least six inches off the floor and away from walls.
- Keep food in original packaging or airtight containers.
- Rotate food correctly.
- Dispose of spoiled food or food that is out of date.
- Never store open cans or bottles in the dry storage area.

Regular inspections will ensure that dry storage areas are pest-free. Correct stock rotation will ensure that older food is used first, and that food pests are spotted. Out-of-date food must be discarded at once. All "best before" date labels should be checked weekly and stock rotated appropriately. Check stock control records weekly.

Covering food prevents contamination from pests, foreign objects, or leaking food. Storing vegetables that are contaminated with soil on the bottom rack will protect other food items in the storage area. Potatoes should be stored in the dark to prevent them from sprouting or turning green.

Food and cleaning chemicals should be stored in separate storage areas to prevent contamination and the tainting of food.

Regular audits of dry storage areas should address temperature, lighting, ventilation, and how staff loads and cleans the storeroom. In particular, check that food is stored correctly, on the appropriate shelves, and that good stock rotation is practiced. Food should be stored away from walls, or stored in mobile, rodent-proof bins. Managers must ensure that food is never left on the floor.

Check less accessible areas for cleanliness and any signs of pest infestation, as well as for the condition of food and food packaging.

Check that food packaging is in good condition and shows no signs of dampness, damage, or pest infestation. Check that cans are not blown, badly dented, holed, rusty, or damaged at the seams. If any defects are found, the cans must be thrown away. Before using canned food, wipe the top with a sanitized clean cloth to prevent dirt from falling into the can.

STORAGE OF SPECIFIC FOODS

While many storage principles apply to all foods, a number apply only to certain foods, and managers must ensure that the proper storage needs of these items are addressed.

Meats

- Meats, such as beef, pork, lamb, veal, and wild game, need to be stored at 41°F (5°C) or below and must be USDA inspected.
- If meat is delivered frozen, it should be placed in the freezer quickly to avoid thawing.
- Store meat in its original packaging, in clean and sanitized containers, or in airtight, moisture-proof wrapping.
- Ground or otherwise non-intact meats must be stored below whole-muscle intact cuts of meat unless they are packaged in a manner that prevents cross-contamination.
- Meats have a short shelf life and, if time/temperature abused, will support rapid bacterial growth.
- Before storing and preparing meat, ensure that it is firm and elastic and has a fresh odor.
- Discard any meat that appears to be spoiled.

Eggs

- Eggs should be stored at 45°F (7°C) or lower.
- Eggs should not be subjected to fluctuating humidity or temperatures, which encourage condensation.
- Don't wash eggs before storing them. Reputable suppliers should wash and sanitize eggs before delivery.
- If egg products are delivered frozen, they should be stored in the freezer quickly to avoid thawing.
- Store non-frozen liquid eggs according to the manufacturer's recommendations.
- Dry egg products should be stored and monitored under dry storage. After dry eggs are mixed with water, they need to be stored at 41°F (5°C) or lower.

Foods with specific storage principles include meats, eggs, poultry, seafood, milk and dairy, fresh fruits and vegetables, and reduced oxygen packaging (ROP) foods.

- It is important to use small amounts when working with eggs so that they are not left out at room temperature.

Poultry

- Poultry includes chicken, turkey, goose, and duck. Poultry should be stored at 41°F (5°C) or below.
- Poultry must be USDA inspected.
- Poultry should be stored in its original packaging, in clean and sanitized containers, or in airtight, moisture-proof wrapping.
- If poultry is delivered frozen, it should be stored in the freezer quickly so that it does not have an opportunity to thaw.
- If poultry is delivered on ice, it can be stored in a refrigerator as is, if it is stored in a self-draining container that is cleaned and sanitized regularly.
- Poultry has a short shelf life and, if time/temperature abused, supports rapid bacterial growth.
- Before storing and preparing poultry, ensure that the flesh is firm and elastic. Poultry can be a source of *Salmonella* and *Campylobacter*.

Seafood

- Seafood includes fish and shellfish (mollusks and crustaceans).
- Fresh fish should be stored at an internal temperature between 32°F and 41°F (0°C and 5°C).
- If fish is delivered frozen, it should be stored in the freezer in a timely manner to avoid thawing.
- Fish fillets and steaks should be stored in their original packaging or securely wrapped in moisture-proof wrapping.
- When fresh, whole fish is delivered on ice, it can be stored in a refrigerator as is, if it is stored in a self-draining container that is cleaned and sanitized regularly. Ice must be changed often.
- Fresh fish should have firm, elastic flesh, clear eyes, and bright red gills. Fresh fish must not have cloudy eyes or a fishy odor.
- Live shellfish should be stored at a temperature of 45°F (7°C) or below.
- Live molluscan shellfish that are on display cannot be for consumption. In order to serve them, there must be approval and variance from a local regulatory agency.
- Smoked fish should be kept in the refrigerator below 36°F (2°C) and consumed within 14 days after smoking. For longer storage, the fish may be frozen immediately after smoking. Store smoked fish in the freezer for no longer than two to three months.

Milk and dairy

- Milk and dairy products, including bakery fillings that contain dairy, must be stored at 41°F (5°C) or lower.
- The use-by or expiration date on milk and dairy products represents the last day the product can be used or sold.
- Fluid milk must be Grade A pasteurized.
- Butters and cheeses must be free of contamination.
- Dairy containers cannot be reused or refilled.

Fresh fruits and vegetables

- Many whole, raw fruits and vegetables, such as carrots and celery, can be refrigerated at 41°F (5°C) or below, at a relative humidity of 85 to 95 percent.
- Citrus fruits, root vegetables, and hard-rind squash can be stored in dry storage.
- The ideal temperature for storing fruits and vegetables in dry storage is 60°F to 70°F (16°C to 21°C).

Smoked fish should be kept in the refrigerator below 36°F (2°C) and consumed within 14 days after smoking. For longer storage, the fish may be frozen immediately after smoking. Store smoked fish in the freezer for no longer than two to three months.

Antonio V. Oquias, 2014/Shutterstock

- Many fruits and vegetables continue to ripen after they are harvested. Fruits and vegetables such as avocados, bananas, and tomatoes ripen best at room temperature.
- Because moisture promotes bacterial growth, most produce should be washed before preparation or service, not upon delivery.
- Fruits and vegetables must be stored away from ready-to-eat foods prior to washing.
- Never store or soak multiple foods or mixed batches of the same food item.

REDUCED OXYGEN FOODS

Reduced oxygen packaging (**ROP**) refers to any packaging procedure that results in a reduced oxygen level.

- Reduced oxygen packaging (ROP) encompasses a large variety of packaging methods, such as controlled atmosphere packaging (CAP) and modified atmosphere packaging (**MAP**), where the internal environment of the package contains less than the normal ambient oxygen level. Using ROP methods in food facilities has the advantage of providing extended shelf life to many foods, as it inhibits spoilage organisms that are typically aerobic. ROP foods should be stored according to the manufacturer's recommendations or at 41°F (5°C) or below.
- While vacuum packaging will slow the growth of microorganisms that require oxygen to grow, it will not prevent microorganisms that do not require oxygen from growing.
- ROP foods are susceptible to *Clostridium botulinum* growth, so discard the product if the package shows signs of contamination, slimy packaging, or an excessive amount of liquid.
- While labeling is important on all packaged foods, it is especially important that ROP foods have labels that list the use-by date, contents, storage temperature, and preparation instructions.
- TCS foods prepared under ROP methods that do not control for growth of toxin formation by *Clostridium botulinum* and *Listeria monocytogenes* require a variance.

KEY TERMS

Ambient temperature The temperature of the surrounding environment.

Clostridium botulinum An anaerobic, intoxication-causing bacteria commonly found in soil and therefore in products that come from soil, such as root vegetables. It can also be found in improperly canned food.

FIFO The acronym used for "first in, first out," which is a system used in stock rotation.

Flow of food The path and direction that food follows through a food facility.

Food safety management system The policies, procedures, practices, controls, and documentation that ensure that food sold by a food business is safe to eat and free from contaminants.

MAP Acronym for modified atmosphere packaging. A physical preservation method that involves changing the proportion of gases normally present around a food item.

Pasteurization The process of applying heat to destroy pathogens.

Perishable foods Those foods subject to spoilage or decay.

ROP Refers to any packaging procedure that results in a reduced oxygen level.

Stock rotation The practice of ensuring the oldest stock is used first and that all stock is used within its shelf life.

Use-by date This is the last date recommended for the use of the product if it is to be used at peak quality. The manufacturer of the product usually determines the date. A retail facility may not sell any food past the use-by date.

ASSESSMENT QUESTIONS

1. Which bacteria are dangerous in ROP foods?
   a. *Bacillus cereus*
   b. *Escherichia coli*
   c. *Clostridium botulinum*
   d. *Staphylococcus aureus*

2. The last day that milk or dairy products can be used or sold in a food facility is the:
   a. Expiration date
   b. Sell-by date
   c. Use by date
   d. Both a and c

3. Which of the following is true regarding food safety and stock rotation?
   a. All TCS and perishable foods should be checked daily.
   b. All TCS and perishable foods should be checked weekly.
   c. All TCS and perishable foods must be used within 14 days.
   d. Food can be used or sold up to three days past the expiration date.

4. Which of the following foods may be stored ABOVE fish in a refrigerator?
   a. Ground beef
   b. Poultry
   c. Whole muscle meat
   d. Fruits and vegetables

5. In order to trace shellfish back to their harvest location, shellfish identification tags must be held on premises for:
   a. Two weeks
   b. 30 days
   c. 60 days
   d. 90 days

6. A significant cause of food contamination during delivery is:
   a. Multiplication of bacteria due to incorrect temperatures
   b. Dirty truck beds
   c. Opened boxes due to shifting during delivery
   d. Poor hygiene of the delivery driver

7. Fresh fish should be stored at what temperature?
   a. between 32°F and 41°F (0°C and 5°C)
   b. between 28°F and 36°F (−2°C and 2°C)
   c. 41°F (5°C) or lower
   d. 45°F (7°C) or lower

8. Powdered egg product that has been mixed with water must be stored at what temperature?
   a. Between 32°F and 41°F (0°C and 5°C)
   b. Between 28°F and 36°F (−2°C and 2°C)
   c. 41°F (5°C) or lower
   d. 45°F (7°C) or lower

9. Food should be stored in what order after it comes off the truck?
   a. Frozen goods, refrigerated goods, dry goods
   b. Refrigerated goods, dry goods, frozen goods
   c. Refrigerated goods, frozen goods, dry goods
   d. Frozen goods, dry goods, refrigerated goods

10. Chemicals must be stored:
    a. On the top shelf of dry storage
    b. On the bottom shelf of dry storage
    c. Underneath the hand sink
    d. In an area away from food

QUESTIONS FOR DISCUSSION

1. Imagine that you are a food manager tasked with hiring a new supplier. What are some qualities that a reputable supplier should possess?

2. When checking in a food delivery on your premises, what are some things you must ensure before accepting the delivery?

3. Food managers must take certain steps to ensure safe dry food storage areas. What are some of those steps?

CHAPTER EIGHT

SAFE FOOD HANDLING

F ood safety managers must ensure that food is handled safely during each operational step through the facility. Every employee who handles food looks to the food manager to know the processes and practices required to ensure food safety. Throughout each of the operational steps—preparation, cooking, cooling, reheating, serving, and packaging—food managers must know what the potential hazards are and how to prevent them.

After reading this chapter, you should be able to:

- Describe the importance of time and temperature controls in food safety.
- List the ways to avoid potential hazards during food preparation.
- Explain how to avoid potential food safety hazards involved in cooking food.

- Describe how to avoid potential hazards involved in cooling and reheating food.
- Explain the potential hazards associated with holding and serving safe food.

TEMPERATURE DANGER ZONE

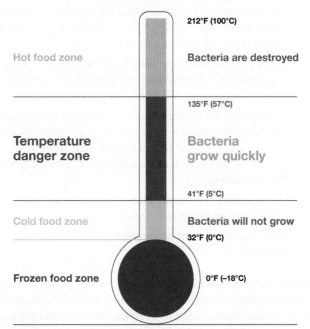

	212°F (100°C)
Hot food zone	**Bacteria are destroyed**
	135°F (57°C)
Temperature danger zone	**Bacteria grow quickly**
	41°F (5°C)
Cold food zone	**Bacteria will not grow**
	32°F (0°C)
Frozen food zone	0°F (–18°C)

Bacteria grow most rapidly in the range of temperatures between 41°F (5°C) and 135°F (57°C), and can double in number in as little as 20 minutes. This temperature range is often called the "Temperature Danger Zone."

Learning Objective: Describe the importance of time and temperature controls in food safety.

Food-poisoning bacteria can multiply to dangerous levels and cause illness when food is left out at room temperature, in the temperature danger zone. The temperature danger zone is any temperature from 41°F to 135°F (5°C to 57°C).

In order to protect food, it is important to minimize the amount of time food spends in the temperature danger zone. For example:

- Raw and cooked perishable foods can spoil quickly and should be refrigerated or frozen promptly.
- Perishable foods standing at room temperature for more than two hours may not be safe to eat.
- Refrigerators should be set at 41°F (5°C) or lower, and freezers should be set at 0°F (–18°C).
- Foods should be cooked long enough and at a high enough temperature to kill the harmful bacteria that cause illness. A meat thermometer should be used to ensure foods are cooked to the appropriate internal temperature:
  - 145°F (63°C) for roasts, steaks, and chops of beef, veal, pork, and lamb
  - 155°F (68°C) for ground beef, veal, pork, and lamb
  - 165°F (74°C) for poultry
- Cold foods should be kept cold, and hot foods should be kept hot.

Refrigerating or freezing food does not necessarily kill bacteria. Once food is brought to room temperature, the bacteria will begin to multiply and become active again. Bacteria also need time at the right temperature in order to multiply to dangerous levels. Food left in the temperature danger zone should be thrown away after four hours.

Time/temperature control for safety (TCS) food is particularly important to monitor. The following procedures will help reduce the chance of time/temperature abuse:

- Ensure that the right thermometers and timers are readily available. Ideally, every kitchen employee should have his or her own thermometer. Thermometers with glass stems must not be used.
- Determine which foods need to be monitored and when. Assign specific staff members the responsibility of monitoring these foods, and require them to keep a record of the foods they monitor.
- Incorporate time/temperature controls into daily practices. For example, require that staff remove only amounts of food that can be prepared in a timely manner from the refrigerator. Use cold food prep items when preparing salads that contain TCS foods, and always cook TCS foods to the required **minimum internal temperatures**.
- Refrigerated TCS foods must be reheated to 165°F (74°C) for 15 seconds within two hours, or be thrown away.
- Set clear policies for employees on the steps to take if food is time/temperature abused.

Some examples of TCS foods are poultry, meat, milk and dairy products, shell eggs, fish, sliced melons and tomatoes, cut leafy greens, sprouts, starchy foods such as baked potatoes, cooked rice and cooked beans, tofu and other soy proteins, and untreated garlic and oil mixtures.

TIME AS A PUBLIC HEALTH CONTROL

Certain foods may be set out for display or set out in an area where temperature control is impossible. In that case, time alone can be used for control.

Time without temperature control can be used as a public health control for up to four hours if:

- The food has an initial temperature of 41°F (5°C) or less when removed from cold holding temperature control, or 135°F (57°C) or greater when removed from **hot holding** temperature control.

- The food is marked or otherwise identified to indicate the time that is exactly four hours after the food is removed from temperature control.

- The food is cooked and served or discarded within four hours from the point in time when the food was removed from temperature control.

- Food in unmarked containers or packages, or marked to exceed a four-hour limit, is discarded.

- A food establishment that serves a highly susceptible population does not use time as specified under this section as the public health control for raw eggs.

Time can be a control in certain circumstances.
Vibe Images, 2014/Shutterstock

THERMOMETERS

Choosing the right thermometer to monitor time and temperature is probably the most effective way to ensure food safety. There are many types of thermometers, and each is designed for a specific task. The thermometer most commonly used in food facilities is a bi-metal stem thermometer. IR (infrared) thermometers and thermocouples are also common. Thermometers with glass sensors or stems cannot be used unless the sensor or stem is encased in a shatterproof coating. The acceptable temperature range for a thermometer used in food preparation areas is 0°F to 220°F (–18°C to 104°C), and there is an allowed variance of +/– 2°F (1°C) for single scale thermometers.

Bi-metal stem thermometers have multiple uses. They are used to measure the temperature of received goods as well as the internal temperature of food. A bi-metal stem thermometer measures temperatures using a metal probe with a sensing area near the tip. It can measure temperatures

Bi-metal stem thermometer

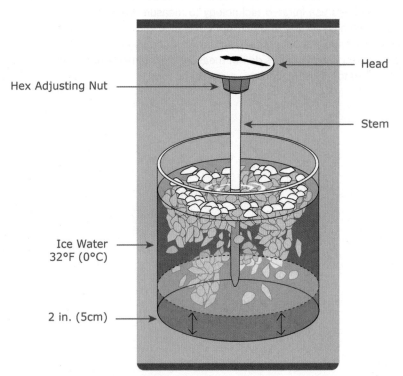

Head

Hex Adjusting Nut

Stem

Ice Water
32°F (0°C)

2 in. (5cm)

Ice water method for thermometer calibration

between 0°F and 220°F (–18°C and 104°C). Before using a bi-metal stem thermometer, make sure it meets the following criteria:

- The temperature markings are clearly numbered and easy to read.
- The accuracy is within 2°F (1°C).
- It has an adjustable calibration nut for accuracy.
- It is designed to undergo cleaning and sanitizing.

Thermocouple
Alexey Stiop/Shutterstock

Thermocouples and thermistors measure temperature using a metal probe with digitally displayed results, and they come in a wide range of sizes. There are other temperature-measuring devices, including immersion probes, surface probes, penetration probes, and air probes. Infrared or laser thermometers use infrared technology to measure the surface temperature of equipment and food, are easy to use, and deliver fast results. Infrared and laser thermometers measure food temperature without contact, so the risk of cross-contamination is nearly eliminated. Infrared and laser thermometers do not measure the internal temperature of food, so they should not be used for cooking, cooling, or reheating food.

| # PREPARATION

Learning Objective: List the ways to avoid potential hazards during food preparation.

The first step in handling food is preparing it for the customer. The main hazards most likely to occur during food preparation are cross-contamination and time/temperature abuse, which both result in the multiplication of bacteria. The following methods outline ways to minimize hazards:

- Prepare raw food and RTE food in different areas. The color coding of equipment will help ensure equipment is not misused.
- Keep hands clean as they are a major source of contamination.
- There should be separate sinks for washing raw foods and for hand washing. Food handlers dealing with raw or TCS foods should wash their hands in separate basins to avoid cross-contamination.
- The waste container for disposable towels should be pedal-operated to avoid cross-contamination due to touching the lid.
- Food handlers should wear gloves to inhibit the spread of bacteria.
- Suitable equipment, such as clean tongs, forks, and cake slicers, should be used to minimize hand contact with food.
- Minimize the amount of food being prepared at a time; batch cooking limits the amount of food that is exposed to hazards. If too much food is prepared, there is a greater risk of bacterial multiplication due to improper holding temperatures.
- There may be a reduction in quality if there is a long gap between preparing and using or serving food. Make sure food is not prepared too far in advance, as this is one of the most common causes of food poisoning.
- Food additives are added to food to lengthen its shelf life or enhance flavor. Additives are also used to alter food so that it does not need time and temperature control.

Since room temperature is within the temperature danger zone, prepared food should immediately be used or refrigerated. By occasionally checking the temperature of food during preparation, a maximum safe time for preparation can be determined. If food is at risk of bacterial multiplication, remedial action, such as changes in preparation procedures or recipes, should be introduced.

Food handlers should work in a logical manner to ensure that work surfaces are kept as tidy as possible. Handlers should effectively clean and sanitize areas and equipment used for preparing food. Follow a "clean as you go" policy, and clean up food spills and waste immediately.

SELF-AUDITS

Regular audits of preparation procedures and activity will make it clear if employees are carrying out their duties safely and following instructions. It will also give an indication as to whether any further training on personal hygiene is needed, or whether introducing any improvements to the workflow could minimize the risk of cross-contamination.

THAWING

Thawing frozen raw food items can be potentially hazardous. Thawing can cause cross-contamination, or the multiplication of spore-forming bacteria that were not destroyed while the food was frozen. The following tips can protect against dangers from thawing:

- Food cannot be thawed on the counter at room temperature. If thawed food is left at room temperature, then dormant bacteria will start to multiply within the food. Even though the center of a food package might still be frozen, the temperature of the outer layer of the food could be within the temperature danger zone.

- Avoid cross-contamination of ready-to-eat foods by thawing raw foods in an area away from ready-to-eat foods. For example, raw frozen chickens cannot be thawed in the same area as cooked food items that are cooling before refrigeration.
- Areas used for thawing should be thoroughly cleaned and sanitized after the food is removed.
- Refrigerate or cook food immediately after thawing.

METHODS FOR THAWING

It is imperative to use a method that will keep food safe as it thaws. Regular audits of the food temperature during the thawing process will determine if safer thawing processes can be introduced, or if staff need further training to ensure an understanding of when thawing is complete. The following are acceptable methods for thawing food.

The refrigerator method

Thaw food in a refrigerator that maintains food at a safe temperature of 41°F (5°C) or less.

The submergence method

Completely submerge frozen food under clean, cold, running water at 70°F (21°C) or below, for a period of time that prevents the temperature of any thawed portion from rising above 41°F (5°C). The water flow must be strong enough to allow particles to float off into an overflow drain. When thawing with water, clean and sanitize the area before and after thawing the food.

Reduced oxygen packaged (ROP) fish that is to be kept frozen until the time of use must be removed from its packaging prior to thawing, or immediately following thawing when using the submergence method.

The submergence method

The cooking process

Food can be thawed as part of the cooking process, provided that it reaches the minimum required internal cooking temperature.

Slacking is the process of gradually increasing the temperature of frozen food from −10°F to 25°F (−23°C to −4°C). Slacking may be used to gradually thaw frozen food, such as shrimp and breaded chicken breasts, in preparation for deep frying.

The microwave method

Thawing food in a microwave oven is acceptable, provided that it is immediately cooked afterward.

Learning Objective: Explain how to avoid potential food safety hazards involved in cooking food.

The main hazards in the cooking or processing stage are: survival of bacteria as a result of inadequate cooking, multiplication of bacteria as a result of prolonged cooking at low temperatures, and contamination. Contamination can occur in many different circumstances. Some examples of how contamination can occur include repeated tasting with the same unwashed spoon, insects falling into uncovered cooking pans, or using equipment made out of inappropriate materials. The following list provides actions and tips for food managers that will ensure a safe cooking process:

Various cooking equipment
monticello, 2014/Shutterstock

- A thermometer can ensure that food is thoroughly cooked. However, if the thermometer is not cleaned and sanitized, it may act as a vehicle of contamination.
- Cooking without using a thermometer to check food's internal temperature cannot guarantee safe food. Food must always be cooked to the minimum internal temperature for the required amount of time.
- Pans and utensils can be vehicles of contamination if they are not thoroughly cleaned between uses. Ensure that pans and utensils are made of suitable materials, are clean, and are free of physical contaminants.
- Do not use copper or aluminum equipment to cook acidic foods as it may result in chemical contamination.
- Sometimes the base of a cooking pan is bigger than the heat source, or stirring is infrequent. In these instances, there will not be an even distribution of heat throughout the food, and some parts of the pan may have undestroyed bacteria that could multiply. Cover pans to avoid physical contamination when not stirring, and ensure that the heat source is under the entire pan base.
- Taste food only with a clean, sanitized spoon. Tasting food with fingers or dirty spoons may result in contamination.

When cooking with a microwave, take appropriate precautions:

- Rotate or stir the food during the cooking process to ensure even heat distribution.
- Ensure that all parts of the food reach a temperature of at least 165°F (74°C).
- Cover the food to maintain moisture.
- Let food sit for two minutes to confirm temperature stability.

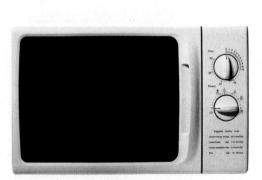

Oleksiy Mark, 2014/Shutterstock

There are monitoring actions that can ensure that cooking controls are working effectively. Regular audits will help to determine if further training is needed for staff on best practices or the use of thermometers. Regularly checking cooking times and internal temperatures of food will confirm that staff is following the correct procedures and instructions.

Check cooking-temperature monitoring records regularly. Once food is prepared, it must be held at the proper temperature until it is served to prevent microorganisms from multiplying and making a customer ill. Hot food must be held at 135°F (57°C) or higher. Cold food must be held at 41°F (5°C) or lower.

COOKING SPECIFIC FOODS

Cooking food to its required minimum internal temperature is the only way to reduce bacteria to a safe level. The minimum internal temperature is not the same for all food items. The minimum internal temperature must be met and held for a certain amount of time. The information presented below covers the minimum internal cooking temperatures for commonly consumed food items.

Eggs	Ground Beef, Pork, other Meats	Beef Steaks	Poultry	Fish	Pork, Veal, Lamb Chops
Raw eggs cooked to order for immediate service must be cooked to at least 145°F (63°C) for 15 seconds. Eggs that will be held for service later must be cooked to 155°F (68°C) for 15 seconds.	Because grinding can spread contaminants throughout the meat, ground meats must be cooked at 155°F (68°C) for 15 seconds.	Steaks must be cooked to at least 145°F (63°C) for 15 seconds.	Poultry must be cooked to at least 165°F (74°C) for 15 seconds.	Fish must be cooked to at least 145°F (63°C) for 15 seconds.	Pork, veal and lamb chops and tenderloin medallions must be cooked to at least 145°F (63°C) for 15 seconds.

Tenderizer-Injected and Mechanically Tenderized Meats	Stuffing and Stuffed Foods	Fruits and Vegetables	Commercially Raised Game and Birds	Leftovers	Ready-to-Eat Foods
Beef and pork injected with tenderizers must be cooked to at least 155°F (68°C) for 15 seconds.	Stuffing made with TCS foods, stuffed fish, stuffed meat, stuffed poultry and stuffed pasta must be cooked to at least 165°F (74°C) for 15 seconds.	Fruits and vegetables being cooked for hot holding should be cooked to at least 135°F (57°C).	Farm-raised game, including elk, deer and bison can follow the same minimum internal temperature guidelines as beef. Steaks should be cooked to 145°F (63°C) for 15 seconds; ground meat to 155°F (68°C) for 15 seconds; stuffed meats to at least 165°F (74°C) for 15 seconds.	Leftovers, or previously cooked TCS foods, must be cooked to at least 165°F (74°C) for 15 seconds.	Ready-to-eat food must be commercially processed in airtight containers or intact packaging from a government-regulated food processing plant. It must be heated to at least 135°F (57°C) within 2 hours.

As consumers often prefer their food cooked rare or medium-rare, food facilities that choose to cook only to those temperatures, and not the recommended temperatures listed here, must include a consumer advisory on their menu. An advisory should state something similar to: "Consuming raw or undercooked meats, poultry, seafood, shellfish, or eggs may increase risk of foodborne illness."

Due to the potential of bacterial illness in a highly susceptible population, the FDA Food Code prohibits offering raw or undercooked **comminuted** (ground, chopped, minced; may be a mixture of two or more types of meat) meat on a children's menu.

Roasts

Roasts, including beef roasts, corned beef, pork roasts, and ham, must be cooked to the following internal temperatures for the specified amount of time:

Temperature	Time (in minutes)
130°F (54°C)	112
131°F (55°C)	89
133°F (56°C)	56
135°F (57°C)	36
136°F (58°C)	28
138°F (59°C)	18
140°F (60°C)	12
142°F (61°C)	8
144°F (62°C)	5
145°F (63°C)	4

Beef or fish (cubed)

Beef or fish cut up into small pieces should be cooked according to these minimum temperatures for the specified amount of time:

Temperature	Time
145°F (63°C)	3 minutes
150°F (66°C)	1 minute
155°F (68°C)	15 seconds
158°F (70°C)	<1 second

NONCONTINUOUS COOKING

Noncontinuous cooking occurs when the initial heating of food is intentionally stopped, cooled, or held prior to sale or service. The food will be completely cooked at a later time, just before service. Raw meat, poultry, seafood, or eggs may be partially cooked during preparation, and later completely cooked just prior to service, if:

1. The initial cooking process is no longer than 60 minutes.
2. The food is immediately cooled after heating and stored in a refrigerator or freezer following TCS food requirements.
3. Prior to sale or service, all parts of the food are reheated to the temperature guidelines above.
4. If the food is not hot held or served immediately, then it must be cooled again following TCS food requirements.

These procedures must be clearly written by the establishment and approved by the local regulatory authority. Documentation should include monitoring procedures and corrective actions, labeling of noncontinuous foods, and assurance that noncontinuous food is stored separately from ready-to-eat foods.

COOLING AND REHEATING

Learning Objective: Describe how to avoid potential hazards involved in cooling and reheating food.

Controlling the processes of cooling and reheating food before serving it to customers is very important to the overall flow of food from purchase to service.

COOLING

The hazards associated with cooling include:

- The multiplication of food poisoning bacteria not destroyed during cooking as a result of inadequate cooking or spore activation
- The contamination of food by bacteria, foreign bodies, or chemicals while cooling

Once food is cooked, it may require storing before being served. Before storing in refrigeration, the food must first cool down safely.

- Cool food quickly after cooking. Cooling food quickly after cooking and refrigerating, or using it as soon as it has cooled, will minimize the period of time that the food is in the temperature danger zone.
- Cooling times depend on the size, thickness, and weight of the food.
- Don't cool hot food in a refrigerator, as this raises the temperature of other foods in the refrigerator and causes condensation.
- Use shallow, stainless-steel pans, as they transfer heat faster than other containers. The pans should be no more than four inches deep and filled to no more than two inches from the top.
- Large amounts of food should be divided if possible.
- Food should be stirred and placed in a **blast chiller** immediately.
- Cool food by placing it in an **ice bath** or steam-jacketed kettle, stirring it with **ice paddles**, or adding ice or cold water as the last preparation step.
- Separate cooling food from raw food, especially raw frozen food that is thawing. If cooling food contacts raw food, it may become contaminated.
- Uncovered food is always at risk of contamination from bacteria, foreign bodies, pests, or chemicals. However, covering food may reduce the speed of cooling.
- Once food is cooled to at least 70°F (21°C), it can be stored in the refrigerator. Be sure there is room for air to circulate around the food, and check the food to guarantee it is cooled to 41°F (5°C) within a total cooling time of six hours.

TCS foods must be cooled:

- From 135°F to 70°F (57°C to 21°C) within two hours
- From 135°F to 41°F (57°C to 5°C) within a total of six hours

Cooling from 135°F to 70°F (57°C to 21°C) is the most critical time, because of the potential for rapid multiplication of bacteria. If food does not get cooled to 70°F (21°C) within two hours, either it must be reheated to 165°F (74°C) and cooled again or it must be discarded. It is vital to cool TCS foods as quickly as possible.

Food managers should regularly audit cooling procedures. These audits will ensure that procedures are safe and that staff follows the processes. Check to be sure that blast chillers are kept in good order and repair. Check that staff follows procedures to minimize cooling times and that food is always at the correct temperature and refrigerated within two hours of cooking.

REHEATING

Food that has been cooled after cooking will eventually be served to customers, either hot or cold. If food is to be served hot, then it will require **reheating**. The reheating of food presents the same potential hazards as cooking: contamination, and the multiplication and survival of bacteria.

- Previously cooked and cooled TCS foods must be rapidly reheated to an internal temperature of 165°F (74°C) for 15 seconds within two hours when being reheated for hot holding.
- Food that does not reach 165°F (74°C) within two hours must be discarded.
- Food that is reheated for immediate service may be served at any temperature as long as it was previously cooked and cooled properly.
- A clean and sanitized thermometer is the best way to ensure food that is thoroughly reheated.
- Food must be reheated with proper food preparation equipment; hot holding equipment must not be used to reheat food.

All parts of the food need to reach an internal temperature of 165°F (74°C) when using a microwave for hot holding. The food must be rotated or stirred, covered, and allowed to stand for two minutes after reheating. A food thermometer can confirm proper temperature throughout the food.

Regularly audit reheating procedures, and check reheating temperatures using a sanitized thermometer. Audits will help to determine whether further training is needed for staff on best practices, correct procedures, personal hygiene, or the use of thermometers. Audits will also allow food managers to check whether better and safer reheating processes can be introduced. Regularly check reheating times, internal temperatures of food, and temperature records to illustrate any major hazards.

If a thermometer is not cleaned and sanitized, it can still pose hazards to thoroughly cooked food.
moreimages/Shutterstock

Learning Objective: Explain the potential hazards associated with holding and serving safe food.

Serving safe food is the ultimate goal of every food manager. The presentation of food to customers is the culmination of everything food handlers and managers work toward.

SERVICE HAZARDS

The hazards associated with the service of food include the multiplication of food-poisoning bacteria because of prolonged periods at room temperature, and contamination from food handlers, equipment, utensils, or price tags. Food left open to self-service is also at risk of contamination from the customer.

- Protect food prepared for self-service by keeping it at the proper temperature.
- If food is served quickly after cooking, cooling, or reheating, it will avoid the multiplication of bacteria. Check food temperatures frequently. Check hot-held and cold-held foods at least every four hours.
- Use thermometers to check food temperatures; don't rely on the holding equipment's temperature gauge. Stir hot food regularly to ensure proper heat distribution.
- If food is on display, then cold food must be kept at or below 41°F (5°C). Ensure that the cold-holding equipment keeps food at this temperature. With the exception of whole fruits and vegetables and cut raw vegetables, food cannot be stored directly on ice.
- Hot food must be kept above 135°F (57°C). The exceptions to this rule are roast beef and pork. If they have been cooked to 130°F (54°C), they can be held at 130°F (54°C).
- There are times when food can be held without temperature controls, but check local regulations before practicing this.
- Set appropriate standards to ensure food is disposed of after a predetermined amount of time. Preparing food in small batches will help minimize the chance of time/temperature abuse and will result in less waste from discarding unconsumed food.
- Customers should not be able to handle open food on display. Food should be prewrapped, sneeze guards should be in place, and there should be plenty of serving utensils to reduce the risk of customer contamination.
- Utensil handles should be longer than the display dishes to prevent them from dropping into food.
- Serve raw food with separate equipment and utensils from ready-to-eat food to prevent cross-contamination. Raw food should be served by different food handlers, and from separate serving counters.
- Food that has been previously served to a customer must not be re-served to a different customer. This includes garnishes, uneaten bread, and unsealed condiments. Generally, only unopened wrapped foods, such as wrapped crackers and condiment packets, can be re-served.
- Ensure that staff understand that ice is a food and should be handled carefully, especially when preparing and serving beverages. Freezing water does not destroy bacteria or toxins, so ice can contain dangerous bacteria or toxins. Only drinking water should be used to make ice.

Safe food is the goal of every food manager.
Kris Vandereycken, 2014/Shutterstock

EQUIPMENT

All equipment and utensils that are likely to come into contact with food should be kept in good condition and should be effectively cleaned and sanitized to prevent cross-contamination and multiplication of bacteria. Train staff to handle equipment and utensils properly. For example, train staff to hold flatware only by the handle, to grip drinking glasses only around the outside, and to not touch the rim or inside edge of glasses.

STOCK AND AUDITS

Except when using ROP methods to package the food, ready-to-eat and TCS foods that are prepared and held in the food establishment for more than 24 hours must be clearly marked with the date by which the food must be eaten, sold, or discarded. This rule applies when food is held at 41°F (5°C) or less for a maximum of seven days. The day of preparation should be counted as day one.

If the establishment sells prepackaged food, check date labels and ensure that stock is correctly rotated. These checks will confirm that service procedures are effective.

The food manager should ensure that regular audits address areas such as:

- Personal hygiene of food handlers
- Cleaning procedures
- Separation of raw and ready-to-eat foods
- Temperature control

KEY TERMS

Blast chiller Rapid cooling refrigeration unit.

Comminuted Reduced in size by grinding, mincing, chopping, or flaking. Ground meats are examples of comminuted food.

Hot holding The storage of cooked food at 135°F (57°C) or higher while awaiting consumption by customers.

Ice bath The method of cooling food in which a container holding hot food is placed into a sink or larger container of ice water. The ice water surrounding the hot food container disperses the heat quickly.

Ice paddles Plastic paddles filled with ice or water and then frozen; they are used to stir hot food to cool it quickly.

Minimum internal temperature The required minimum temperature that the internal portion of food must reach to sufficiently reduce the number of microorganisms that might be present. This temperature is specific to the type of food being cooked.

Noncontinuous cooking Occurs when the initial heating of food is intentionally stopped, cooled, or held prior to sale or service.

Reheating The process of heating previously cooked and cooled foods to a temperature of at least 165°F (74°C).

Slacking The process of gradually increasing the temperature of frozen food from −10°F to 25°F (−23°C to −4°C); generally in preparation for deep frying.

ASSESSMENT QUESTIONS

1. The process that can be used for slowly thawing food in preparation for deep frying is called:
   a. Serving
   b. Slacking
   c. Sloughing
   d. Submersion

2. Throw out leftover food held at _____ or lower after seven days.
   a. 32°F (0°C)
   b. 41°F (5°C)
   c. 70°F (21°C)
   d. 100°F (38°C)

3. Poultry needs to be cooked to a minimum internal temperature of:
   a. 135°F (57°C)
   b. 145°F (63°C)
   c. 155°F (68°C)
   d. 165°F (74°C)

4. Stuffing and stuffed meat needs to be cooked to a minimum internal temperature of:
   a. 135°F (57°C)
   b. 145°F (63°C)
   c. 155°F (68°C)
   d. 165°F (74°C)

5. Commercially processed, ready-to-eat food needs to be reheated to:
   a. 212°F (100°C)
   b. 165°F (74°C)
   c. 135°F (57°C)
   d. 70°F (21°C)

6. Substances added to food to lengthen its shelf life or to alter food so that it does not need time and temperature control are called:
   a. Stuffing
   b. Coolers
   c. Food additives
   d. Hazards

7. All hot foods can be held at:
   a. 115°F (46°C)
   b. 135°F (57°C)
   c. 155°F (68°C)
   d. All hot foods need to be held at their minimum internal temperature.

8. A quick way to cool a large amount of food is:
   a. Dividing up the food
   b. Using a blast chiller
   c. Stirring with ice paddles
   d. All of the above

9. TCS stands for:
   a. Time/temperature control for safety
   b. Time/temperature cost for security
   c. Treatment of coolers for sanitation
   d. Temperature that could spoil food

10. An example of a TCS food is:
    a. Oats
    b. Dried basil
    c. Sliced melon
    d. Sliced bread

QUESTIONS FOR DISCUSSION

1. What are some procedures you can follow to help reduce the chance of time/temperature abuse in your food service establishment?

2. Imagine you are the manager of a busy seafood restaurant. You are tasked with hosting a training session on the proper thawing procedures for 10 employees. In preparation for your session, what are the key points of safe thawing? What are the biggest hazards of thawing?

3. What are some ways that food can be contaminated during cooking? How can you prevent contamination during cooking?

4. Marcus works in the kitchen at a popular Mexican restaurant. He has just cooked five pounds of chicken for dinner service. How can Marcus safely cool and store the cooked chicken?

5. Your manager is hosting a training session on serving safe food. In preparation for the session, she asks you to write down three hazards associated with food service. What do you write?

THE HACCP APPROACH TO FOOD SAFETY

In 1993, the FDA combined three previous codes on food safety to develop the first comprehensive Model Food Code. Within this new singular document based on scientific evidence, the FDA included Hazard Analysis and Critical Control Point (commonly referred to as **HACCP**—pronounced *HA-sip*) principles. The HACCP approach is a food safety management system that focuses on hazard control throughout the food flow rather than on sanitation alone. HACCP food safety programs are mandatory for most food manufacturers and schools. Some states already conduct HACCP-based inspections in retail food facilities. This trend will continue, because many regulators consider the HACCP approach to be the best defense of the food chain "from farm to table."

The HACCP approach began with NASA's Apollo space program. A project of the Pillsbury Company, the original goal of the **HACCP plan** was to ensure that astronauts had safe food. The consequences of astronauts suffering from diarrhea or vomiting in a zero gravity environment would be catastrophic, and could lead to mission failure.

Most reputable food establishments were aware of potential food safety problems and implemented appropriate controls before HACCP; however, the food industry made very little attempt to determine which controls were most important to reduce the risk of foodborne illness. Instead, food safety control relied heavily on end-product testing, which had many disadvantages, such as:

- Control was reactive, and action was taken after the problem occurred.
- Testing was often slow, and results would take considerable expertise to interpret. The delay sometimes meant that sometimes the product reached consumers before a problem was diagnosed.

- The cost of sampling and analysis was high.
- Only a few staff members were directly involved in food safety.

After reading this chapter, you should be able to:

- Explain the purpose and principles of the HACCP approach to food safety.

- Explain the hazard analysis process.

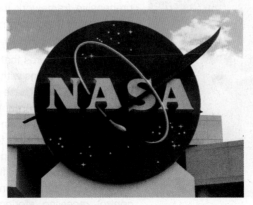

HACCP began with NASA.
L Galbraith/Shutterstock

Learning Objective: Explain the purpose and principles of the HACCP approach to food safety.

To effectively reduce the occurrence of foodborne illness risk factors, operators of retail and food service establishments must focus their efforts on achieving active managerial control. The term **active managerial control** is used to describe industry's responsibility for developing and implementing food safety management systems that prevent, eliminate, or reduce the occurrence of foodborne illness risk factors. Active managerial control embodies a preventive rather than a reactive approach to food safety through a continuous system of monitoring and verification.

Many tools can be used to provide active managerial control of foodborne illness risk factors. Regulatory inspections and follow-up activities must be proactive and use an inspection process designed to assess the degree of active managerial control that retail and food service operators have over foodborne illness risk factors. Regulators should assist operators in developing and implementing voluntary strategies to strengthen existing systems that prevent foodborne illness risk factors.

FOOD MANAGEMENT SYSTEMS

Elements of an effective food safety management system may include the following:

- Certified food protection managers (CPFM) who have shown a proficiency in required information by passing a test that is part of an accredited program
- Standard operating procedures (SOPs) for performing critical operational steps in a food preparation process, such as cooling and reheating
- Recipe cards that contain the specific steps for preparing a food item and the food safety critical limits that need to be monitored and verified; for example, final cooking temperatures
- Purchase specifications
- Design and maintenance of the equipment and facility
- Monitoring procedures
- Recordkeeping
- Employee health policy for restricting or excluding ill employees
- Manager and employee training
- Ongoing quality control and assurance
- Specific goal-oriented plans, such as risk control plans (RCPs), that outline procedures for controlling foodborne illness risk factors

A food safety management system based on Hazard Analysis and Critical Control Point (HACCP) principles contains many of these elements. HACCP programs provide a comprehensive framework for a food facility operator to effectively control the occurrence of foodborne illness risk factors.

In order for a HACCP system to be effective, a strong foundation of procedures that address the basic operational and sanitation conditions within a facility must be developed and implemented first. These procedures are collectively termed **prerequisite programs**. When prerequisite programs are in place, more attention can be given to controlling hazards associated with food and food preparation. Prerequisite programs may include things such as facility design, supplier control, ingredient specifications, equipment design, cleaning and sanitation, personal hygiene, employee training, pest control, receiving, storing, and shipping procedures, traceability, and recall procedures.

Active managerial control reduces foodborne illness.
jdwfoto/Shutterstock

PURPOSE AND PRINCIPLES

What is unique about the HACCP approach as a food safety management program is that it is proactive rather than reactive. In other words, a HACCP program tries to prevent food-borne illnesses before they occur instead of dealing with the issues that caused illness after the fact.

The HACCP approach examines each step of the food-handling process and identifies potential hazards. Identifying hazards in advance, and controlling them, will help prevent them from ever becoming an end-product problem for consumers.

A HACCP program takes time to implement and should not implemented too quickly or haphazardly for the benefit of an inspector. It requires mapping and examining even the most mundane food-handling processes. Once it is properly implemented, the HACCP approach can help ensure consistently high standards of food safety every day.

The seven steps of building a HACCP plan:

1. Conduct a **hazard analysis** to identify the hazards across the process or operation and specify control measures, actions required to prevent or eliminate a food safety hazard to reduce the hazard to an acceptable level.

2. Determine **critical control points** (CCPs) to pinpoint which of the steps where hazards were identified are critical to food safety.

3. Establish **critical limits**, target levels, and tolerances for each **CCP**.

4. Establish a **monitoring** system for each CCP, through scheduled testing or observations.

5. Establish **corrective actions** to be taken when a CCP is out of control; that is, when a critical limit is breached.

6. Establish **verification** procedures, which include appropriate **validation**, together with a review to confirm that the HACCP program is working effectively.

7. Establish documentation and recordkeeping for all procedures relevant to these principles and their application.

When conducting a hazard analysis, food manufacturers usually use food commodities as an organizational tool and follow the flow of each product. This is a very useful approach for producers or processors, as they are usually handling one product at a time. By contrast, in retail and food service operations, foods of all types are produced together to create the final product. This makes a different approach to hazard analysis necessary. Conducting the hazard analysis by using the food preparation processes common to a specific operation is often more efficient and useful for retail and food service operators. This is called the process approach to HACCP.

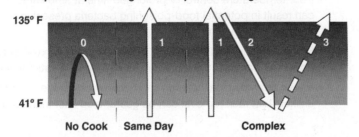

Complete Trips Through the Temperature Danger Zone

135° F

41° F

No Cook   Same Day   Complex

The three food preparation process categories are based on the number of times food passes through the temperature danger zone.

THE PROCESS APPROACH TO HAACP

The process approach to HACCP can best be described as dividing the food preparation methods in a facility into categories based on the number of times a specific food will be in the temperature danger zone, then analyzing the hazards and placing managerial controls on each grouping.

PROCESS 1: Food Preparation with No Cook Step

Example flow: Receive-Store-Prepare-Hold-Serve (Other food flows are included in this process, but there is no cook step to destroy pathogens.) Example product: tossed salad. Possible control: Keep product at a temperature of 41°F (5°C).

PROCESS 2: Preparation for Same Day Service

Example flow: Receive-Store-Prepare-Cook-Hold-Serve (Other food flows are included in this process, but there is only one time the product is in the temperature danger zone.) Example product: a hamburger. Possible control: Maintain product temperature at 135°F (57°C) until service. Be sure that hamburger is first cooked to 155°F (68°C).

PROCESS 3: Complex Food Preparation

Example flow: Receive-Store-Prepare-Cook-Cool-Reheat-Hot Hold-Serve (Other food flows are included in this process, but there are always two or more times that the product is in the temperature danger zone.) Example product: chili that will last a few days. Possible control: Ensure correct product cooling, reheating, and holding procedures.

Learning Objective: Explain the process of developing a HACCP plan.

Building a food safety management system takes time, patience, and determination. Careful consideration must be given to all aspects of an operation affecting food safety. To assist in building a food safety management system, a series of steps (the seven HACCP principles) were developed as a guide through the process. Each of the seven HACCP principles are addressed here:

1. CONDUCTING A HAZARD ANALYSIS

The goal of this type of analysis is to identify the hazards across a process or operation, and then to specify control measures. **Control measures** are the actions required to prevent or eliminate a food safety hazard or reduce it to an acceptable level.

All other HAACP principles depend on the **hazard analysis**. The choices made during hazard analysis determine the scope and success of the HACCP program. A hazard analysis consists of six steps:

- **STEP ONE:** Assemble a team of staff members to conduct the analysis.

  Team members should be familiar with the selected products and processes. It's important for the food manager to not just be part of the team but also to support and lead it. Management should ensure that all team members have the time, resources, and training they need to complete the analysis.

- **STEP TWO:** Determine which products and processes need to be examined.

  Examples of processes to examine include:

  - Perishable raw food that is cooked and served hot

  - TCS food that is served cold

  - TCS food that is served hot

  - Frozen raw food that is cooked, cooled, and served cold

- **STEP THREE:** Prepare flow diagrams of any related processes.

  A flow diagram is a map of the sequence of steps or operations involved with a particular food item or process, usually from receipt of raw materials to the consumer. A flow chart of the kitchen—from the receiving area to the service area—and how food moves through the area can assist in seeing where contamination might occur. Team members can then validate the flow diagrams to make sure they accurately reflect what happens in actual practice.

- **STEP FOUR:** Brainstorm potential hazards.

  Work with the HACCP team through each step of the flow diagram and determine which hazards could occur at specific points. Potential food hazards include, but are not limited to, the following:

  - Poor temperature control or prolonged time at ambient temperatures can result in potential food-poisoning bacteria present in food to multiply to large numbers.

  - Failure to cook food thoroughly can result in the survival of food-poisoning bacteria, which can make anyone who eats the food ill.

  - Physical or chemical hazards can occur at any stage in the process. It's unlikely that their removal can be guaranteed later in the process.

Examine areas of potential contamination.

dmytro herasymeniuk/Shutterstock

HACCP Flowchart

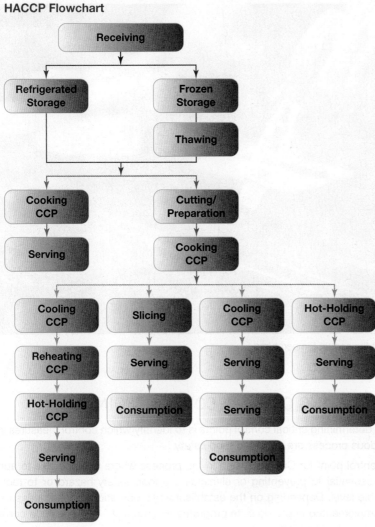

Use food preparation processes to conduct hazard analysis.

- **STEP FIVE:** Perform a **risk assessment** to determine the significance of potential hazards.

  Whether a hazard is considered significant depends on the likelihood of the hazard occurring and the seriousness of potential consequences, or **risk**. At this point, low-risk hazards that don't have severe consequences are set aside and excluded from the hazard analysis. Such hazards are best addressed as quality control issues.

- **STEP SIX:** Determine control measures for high-risk hazards.

  Control measures are actions required to prevent or eliminate a food safety hazard or to reduce the hazard to an acceptable level. Controls can be general or specific. General controls relate to initial best practices at the beginning of the process, such as approved suppliers, effective cleaning and sanitizing, pest management, staff supervision and training, good design, and effective maintenance. Specific controls include monitoring items such as the time and temperature of cooking, or the volume or weight of preservatives added to a food product.

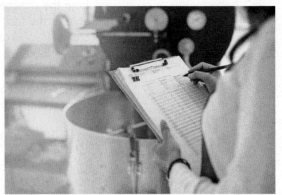

Conducting a hazard analysis identifies where hazards can occur.

Mila Supinskaya/Shutterstock

Keep food at the correct temperatures.
moreimages/Shutterstock

2. DETERMINING CRITICAL CONTROL POINTS

The goal of determining critical control points is to identify which control measures in a potentially hazardous process are critical to food safety.

A critical control point (or CCP) is a step in the process where it is possible to apply a control that is essential to preventing or eliminating a food safety hazard or to reducing it to an acceptable level. Depending on the establishment's operation, control measures may be effectively implemented in prerequisite programs. Examples of CCPs include cooking, cooling, and hot and cold holding.

3. ESTABLISHING CRITICAL LIMITS

The goal of establishing critical limits is to identify **target levels** and tolerances for each CCP.

Critical limits are the values of monitored actions at CCPs that separate the acceptable from the unacceptable. Critical limits must be measurable. The results must be obtainable at the facility without sending samples away for laboratory analysis. Critical limits should involve physical characteristics such as temperature, time, pH, and the weight and size of food.

Target levels are specified values for control measures that eliminate or reduce hazards at CCPs. They act as a buffer zone, providing a tolerance for operation that can prevent a failure to meet critical limits.

4. ESTABLISHING A MONITORING SYSTEM

The goal of establishing a monitoring system is to create a testing and observation schedule for each CCP.

Monitoring is the planned observation and measurement of control parameters to confirm that processes are under control and that critical limits are not exceeded. Monitoring is required to identify breaches of target levels, to identify deviations and trigger corrective actions, to provide records for verification, to investigate complaints, or for due diligence. Monitoring can be manual, automatic, continuous, or at set frequencies. Most important, monitoring must permit rapid detection and correction, regardless of the type of observation or measurement used.

Critical Control Point Decision Tree

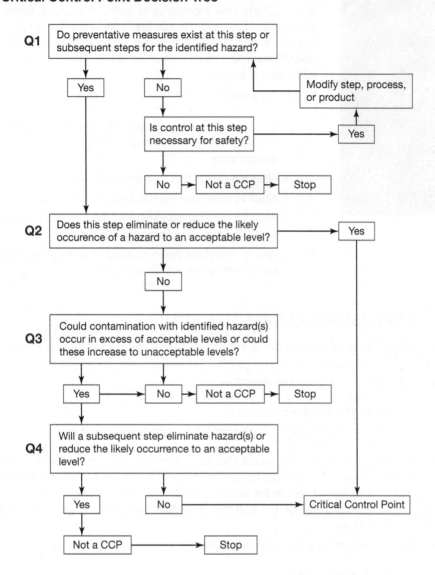

Q1 — Do preventative measures exist at this step or subsequent steps for the identified hazard?

- Yes
- No → Is control at this step necessary for safety?
  - Yes → Modify step, process, or product
  - No → Not a CCP → Stop

Q2 — Does this step eliminate or reduce the likely occurence of a hazard to an acceptable level?

- Yes
- No

Q3 — Could contamination with identified hazard(s) occur in excess of acceptable levels or could these increase to unacceptable levels?

- Yes
- No → Not a CCP → Stop

Q4 — Will a subsequent step eliminate hazard(s) or reduce the likely occurrence to an acceptable level?

- Yes → Not a CCP → Stop
- No → Critical Control Point

MONITORING METHODS

Observations

The appearance, smell, texture, and other physical characteristics of food are valuable for obtaining a rapid assessment of food standards. For example, food that smells stale or musty indicates a problem. The corrective action would be to discard the food.

Measurements

Measurements include checking temperatures, times, and pH values; or checking goods receipt records, cleaning schedules, and maintenance records.

Audits

An audit is a systematic, independent, and documented process for verifying proper food safety practices within a facility. Audits may address the entire operation or focus on specific procedures and practices. Conducting regular self-audits will help confirm that the food safety procedures are working as planned.

Rotten food must be discarded.

StepanPopov/Shutterstock

Monitoring temperature
James "BO" Insogna, 2014/Shutterstock

To ensure proper food safety all the time, the entire staff must know and understand the importance of CCPs, monitoring, target values, and critical limits, and the role they play in monitoring procedures. Specific training may be required. All food handlers, not just the HACCP team, should be competent in HACCP procedures as needed for their activities.

Monitoring systems should state:

- Who is responsible for monitoring and who is responsible for checking that monitoring has been carried out appropriately
- What the critical limits, target levels, and tolerances are
- When to undertake the monitoring
- Where in the flow diagram monitoring should be done—at the CCP or as close as possible to the CCP
- How to undertake the monitoring, including details of what equipment to use and its calibration

5. ESTABLISHING CORRECTIVE ACTIONS

The goal of establishing corrective actions is to identify the necessary procedures to take when a critical limit is not met.

Corrective action usually has two objectives:

- Dealing with any affected product
- Reviewing the CCP and bringing the process back under control

There are several ways a food product that tests outside critical limits can be treated. For example, a hazardous product may be quarantined and subjected to further testing. A product's shelf life may be reduced or the product may be destroyed. Sometimes a process that contributed to a hazard may be adjusted and corrected—by extending the cooking time of a product, for example.

Document procedures for each corrective action. Procedures should specify:

- The action(s) to take
- The person responsible for taking action and a clear chain of command
- Who should be notified
- Who is authorized to stop and then restart production or sales; this will normally be management
- The treatment of affected products

6. ESTABLISHING VERIFICATION PROCEDURES

The goal of establishing verification procedures is to put a system in place to confirm that a HACCP program is working.

Verification involves methods, procedures, and tests, in addition to those used in monitoring, to determine compliance with the HACCP plan.

Verification comprises three stages:

1. Validation
2. Ensuring a satisfactory HACCP system
3. Review

Validation involves obtaining evidence that elements of the HACCP plan are effective, especially CCPs and critical limits.

Review plan
Pressmaster, 2014/Shutterstock

Ensuring that the HACCP system as a whole is effective usually involves self-auditing against the HACCP plan to ensure correct implementation. Auditing is important to guarantee that:

- The food flow diagram remains valid.
- Hazards are being controlled.
- Monitoring is satisfactory.
- When necessary, appropriate corrective action is taken.

The final stage of verification is review. The HACCP plan should be reviewed at least annually to ensure that it remains effective. Remember that the review results must always be recorded.

7. ESTABLISHING A RECORDKEEPING SYSTEM

To guarantee the success of any HACCP program, a management system needs to be in place to keep HACCP documentation and records up to date and effective.

The simplest, most effective recordkeeping system that integrates well with the operation should be used. The amount and type of paperwork required to support HACCP systems should be proportionate to the type and size of the food business and the risks involved with processes. Accurate records and documentation are essential for verification and auditing. Records and documentation are also useful for demonstrating how food safety is managed. It is not uncommon for inspectors to review records and documentation. Records are also vital when investigating customer complaints and claims of alleged food poisoning.

Recordkeeping
jdwfoto/Shutterstock

The HACCP plan is the principle document in any HACCP program. Program documentation should include the following items:

- Details of the HACCP team and responsibilities of team members
- Hazard analysis, including product and process descriptions
- Flow diagram and CCP determination
- Critical limits, target values, and corrective actions
- Monitoring and recordkeeping procedures
- Verification procedures

Other documents that show how the HACCP plan has been supported include product specifications, a floor or room plan, and details of prerequisite programs such as personal hygiene and pest management programs.

Examples of useful records are:

- Modifications to the HACCP system, including any details or notes from the review
- Audit reports
- Customer complaints
- Calibration of instruments
- Approved supplier list
- Stock rotation records
- Staff health records
- Cleaning schedules and training records

Monitoring records should be signed and dated by the food handler who does the monitoring and countersigned by that person's manager.

A food manager's responsibilities concerning the HACCP system include:

- Becoming familiar with the HACCP principles
- Determining whether the facility is subject to regulations based on the HACCP principles
- Considering how the facility would benefit from an HACCP program

KEY TERMS

Active managerial control (AMC) The purposeful incorporation of specific actions or procedures by industry management into the operation of their businesses in order to attain control over foodborne illness risk factors.

Control measures The actions required to prevent or eliminate a food safety hazard or reduce it to an acceptable level.

Corrective action The action to be taken when a critical limit is breached.

Critical control point (CCP) A point or procedure in a food system where control can be applied and is essential to prevent or eliminate a food safety hazard or reduce it to an acceptable level.

Critical limit The value of a monitored action that separates the acceptable from the unacceptable.

HACCP Acronym for hazard analysis and critical control point. A food safety management system that identifies, evaluates, and controls hazards that are significant for food safety.

HACCP plan Documentation completed during the HACCP study and implementation. It includes the hazard analysis, the flow diagram, the HACCP control charts, monitoring records, verification details, and modifications to the system.

Hazard analysis The process of collecting and evaluating information on hazards and on conditions leading to their presence to decide which are significant for food safety and therefore should be addressed.

Monitoring The planned observation and measurement of control parameters to confirm that processes are under control and that critical limits are not exceeded.

Prerequisite programs Procedures, including standard operating procedures (SOPs), that address basic operational and sanitation conditions in an establishment.

Risk The likelihood of a hazard occurring.

Risk assessment The process of identifying hazards, assessing risks and severity, and evaluating their significance.

Target level The predetermined value for the control measure that will eliminate or control the hazard at a control point.

Validation The element of verification focused on collecting and evaluating scientific and technical information to determine if the process or procedure will effectively control the hazards.

Verification The application of methods, procedures, and tests to determine compliance with a food safety plan, in addition to those used in monitoring to determine compliance with a HACCP plan.

ASSESSMENT QUESTIONS

1. What is the first step to building a HACCP plan?
   a. Conduct a hazard analysis to identify the hazards across the process or operation and specify control measures.
   b. Establish corrective actions to be taken when a CCP is out of control; that is, when a critical limit is breached.
   c. Establish documentation and recordkeeping for all procedures relevant to these principles and their application.
   d. Establish a monitoring system for each CCP, through scheduled testing or observations.

2. The term *active managerial control* is used to describe:
   a. A hands-on food safety manager
   b. The industry's responsibility for developing and implementing food safety management systems that prevent, eliminate, or reduce the occurrence of foodborne illness risk factors
   c. A disciplinary measure taken when employees do not respond to managerial supervision
   d. A managerial style that is most effective in large-scale food service operations

3. What is the primary goal of establishing critical limits?
   a. Adhering to cold-cooking temperatures
   b. Establishing a warning system for employee uniform infractions
   c. Identifying target levels and tolerances for each CCP
   d. Clearing the kitchen of potentially hazardous toxins

4. What are control measures?
   a. Actions required to prevent or eliminate a food safety hazard or to reduce the hazard to an acceptable level
   b. Actions taken to monitor employee attendance records
   c. Actions that are contradictory to a satisfactory HACCP system
   d. Final cooking temperatures

5. The HACCP approach began with:
   a. McDonald's
   b. University of Michigan
   c. NASA's Apollo space program
   d. Habitat for Humanity

6. TCS food must be served:
   a. By trained employees
   b. In steel containers only
   c. Without dairy
   d. At the correct temperature

7. What is the goal of establishing a monitoring system?
   a. To reduce the number of health inspector visits per year
   b. To create a testing and observation schedule for each CCP
   c. To adhere to a regular maintenance schedule
   d. To boost employee morale

8. Which of the following is NOT one of the stages of verification?
   a. Ensuring a satisfactory HACCP system
   b. Validation
   c. Review
   d. Authentication

9. Obtaining evidence that shows a procedure is effective is called:
   a. Verification
   b. Validation
   c. Establishment
   d. Corrective action

10. HACCP program records should include each of the following EXCEPT:
    a. Temperature logs
    b. Training logs
    c. Delivery logs
    d. Mileage logs

QUESTIONS FOR DISCUSSION

1. What are some elements of an effective food safety management system? In your opinion, which is the most important element?

2. Imagine you are the manager of a popular seafood chain restaurant and you receive an e-mail from the company's corporate headquarters. As part of refresher training, you are asked to form a hazard analysis plan. Be sure to explain all six steps in detail.

FOOD SAFETY STANDARDS

Food manager training teaches food safety principles such as cleaning and sanitizing, separating raw and ready-to-eat foods, cooking foods thoroughly, and cooling foods correctly so that food consumers stay safe.

Standards are necessary to ensure consistency, and to provide a reference point to determine when a target is achieved. Standards are also needed to determine that a task, such as cleaning a meat slicer, has been completed properly. Standards can relate to facility, people, or products and may be used in several different ways. There are a variety of standards, including voluntary standards such as in-house standards, and legal standards such as federal, state, and local regulations. Many individuals and organizations can set standards, including the manager, company, customer, government, trade unions, independent standards authorities, professional associations, local authorities, and food safety authorities.

A company's standards should be clearly set out in the food safety policy. The **food safety policy** outlines management's responsibilities and communicates standards to staff. The policy is effectively a commitment to produce safe food, provide satisfactory facilities and equipment, ensure that legal responsibilities are met, and maintain appropriate records.

A food safety policy should include procedures for:

- Approval and monitoring of suppliers and product delivery
- Personal hygiene, health screening, and reporting of illnesses
- Informing and supervising staff
- Implementing an effective training program
- Effective temperature control and monitoring practices
- Integrated pest management

- Cleaning and sanitizing, including cleaning schedules
- Waste management
- Customer complaints
- Foodborne illness outbreaks
- Information on food allergies
- Quality control and assurance
- Handling inspections and inspectors

After reading this chapter, you should be able to:

- Explain the purpose and application of the FDA Food Code.
- Explain the purpose of an inspection program.
- Identify best practices for sampling foods in a food facility.

- Explain the importance of using dates and labels on food.
- Describe proper safety precautions for hazardous materials.

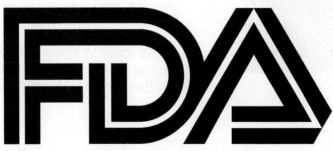

Courtesy of The Food & Drug Administration

Learning Objective: Explain the purpose and application of the FDA Food Code.

The FDA Model Food Code represents the federal government's "best advice" for minimizing the incidence of foodborne illness. It is not a law; however, regulatory authorities at all levels of government typically use the Food Code to develop or update their food safety rules. In fact, according to the Association of Food and Drug Officials, 100 percent of U.S. states and territories have implemented some form of the FDA Food Code (1993–2013) as of June 2015.[1] If a food manager transfers to a restaurant or food facility in another state, the Food Code will still apply, even if state or local regulations change.

HOW THE FDA FOOD CODE IS DEVELOPED

The Food and Drug Administration (FDA) issues the Food Code every four years and then amends it every two years with a supplement. While the FDA is listed as the author of the Food Code, the FDA isn't the only group involved in creating it. Other federal agencies, mainly the USDA and the CDC, work with the FDA to create the Food Code.

One group outside the government that has input into the contents of the Food Code is the Conference for Food Protection (CFP). CFP is a nonprofit organization that was formed in 1971 to create an equitable partnership among food industry regulators, professionals, academics, and consumers. The conference takes place every two years before the release of a revision or supplement to the FDA Food Code. The goal of the conference is to identify problems, formulate recommendations, and develop practices to ensure food safety. After the conference, the CFP presents its findings to FDA. The FDA takes the CFP findings into consideration while developing the Food Code.

The current FDA Food Code includes hundreds of pages of recommendations that cover the following topics:

CFPM

- Management and personnel (including supervision, employee health, personal cleanliness, and hygienic practices)
- Food preparation and handling (including criteria for receiving, storing, displaying, handling, preparing, serving, and transporting)
- Equipment, utensils, and linens (including design, construction, operation, maintenance, repair, cleaning and sanitizing procedures, and protection)
- Water, plumbing, and waste (including utilities and services)
- Physical facilities (including design, construction, maintenance, repair, capacities, location, and operation)
- Poisonous or toxic materials (including identification and labeling, storage, sale, and application)
- Compliance and enforcement (including code applicability, plan submission and approval, permit to operate, inspection, and correction of violations)

States that have adopted the most recent Food Code now require that at least one food facility employee with management or supervisory responsibility be a certified food protection manager (CFPM).

LESSON 2 | INSPECTIONS

Learning Objective: Explain the purpose of an inspection program.

The FDA Food Code defines the **person in charge (PIC)** as: "the individual present at a food establishment who is responsible for the operation at the time of inspection." In many cases, the food manager will be the person in charge. However, when the food manager is not present at the food facility, it's important to have identified a person in charge to oversee safe food service and to deal with questions or concerns regarding food safety. In addition to ensuring that the facility is serving safe food, the person in charge must demonstrate a wide range of knowledge both in day-to-day operations and during an inspection.

INSPECTIONS

The primary purpose of a food service inspection program is to protect the public's health by determining if a facility provides food that is safe, unadulterated, and honestly presented. Inspection is the principle tool a regulatory agency has for detecting procedures and practices that might be hazardous. Inspections can also identify actions to correct deficiencies.

Serving unsafe food can jeopardize the health of customers and the establishment. Inspections can show how well the food facility is doing when it comes to following standard food safety practices. An excellent food safety inspection report is one of the most effective forms of advertisement for the facility, because inspection data is public information. Inspection data is readily available on the Internet in most jurisdictions.

Any establishment that serves food to the public is subject to inspection by one or more regulatory authorities. This is true even if there is no charge for the food or if consumption of the food takes place off-site.

While inspection frequency varies by jurisdiction, some regulatory authorities inspect all food establishments on the same schedule. Other authorities inspect high-risk food establishments more frequently.

Factors used to determine risk include:

- **Establishment size.** Larger establishments employ more people and serve more customers, and subsequently have an increased risk of foodborne illness.
- **Foods served.** Extensive handling of raw foods increases the possibility of foodborne illness.

Inspector
XiXinXing/Shutterstock

- **Susceptibility of clientele.** Children, the elderly, and people with suppressed immune systems are more susceptible to foodborne illness.
- **Previous compliance history.** Establishments with repeat violations are typically considered high risk until they achieve a consistent record of compliance.

Some inspectors provide a notice in advance for an inspection. This gives food managers the opportunity to prepare mentally and logistically for an inspection. However, most inspectors arrive unannounced during the hours of operation and ask for either the manager or the person in charge.

A facility may be subject to any one of three types of inspections—traditional, risk-based, or HACCP-based. It's not uncommon for regulatory authorities to employ combinations of these three inspection types.

Traditional inspections

A traditional inspection ends with a number or letter grade. Most jurisdictions use a number-based system for scoring, and then assign a letter grade for the benefit of the public.

- For each violation, an inspector may subtract from one to five points:
  - One or two points are subtracted for noncritical violations.
  - Four or five points are subtracted for critical violations.
- Noncritical violations must be addressed before the next routine inspection.
- There is a time limit—usually 48 hours or less—to correct critical violations.
- Failure to correct violations may result in fines or closure.

Many regulators, including the FDA, find the total scoring aspect of traditional inspections problematic. Although inspectors subtract points for each broad violation category, they don't subtract points for each instance of a violation. This means that a facility with multiple violations in one category could receive the same inspection score as a facility that only had a single violation in that category.

Risk-based inspections

Because of what the FDA views as limitations of traditional inspection systems, the FDA now recommends the use of risk-based inspections. Risk-based inspections make a distinction between priority items, priority foundation items, and core items.

In a risk-based inspection, the application of a **priority item** contributes directly to the elimination, prevention, or reduction of hazards associated with foodborne illness or injury. These items include things such as cooking, reheating, cooling, and hand washing. Per the Food Code, a violation of a priority item must be corrected within 72 hours.

Priority foundation items support, facilitate, or enable priority items. Violations of priority foundation items require specific actions or procedures to attain control of a hazard. Violations of priority foundation items must be corrected within 10 days.

Core items are those that don't have a direct impact on food safety. These include factors such as general physical facility conditions and general work practices. Instead of deducting one to five points for a violation, the risk-based inspectors assess an establishment as "in" or "out" of compliance for one or more risk factors. Violations of core items must be corrected in 90 days, however, the inspector will set a time limit for the correction of all violations in his or her jurisdiction. If the food facility is still in violation of any regulation when the time limit has expired, the facility may be subject to fines or closure.

HACCP-based inspections

HACCP-based inspections focus on the control of hazards throughout the food flow. During this complex type of inspection, inspectors observe and assess how an establishment receives, stores, prepares, and serves food.

HACCP-based inspections don't use a traditional scoring system. As with risk-based inspections, the main goal of HACCP-based inspections is to determine and resolve violations of priority and priority foundation items.

Regular self-audits can help prepare a facility for an inspection.

Tyler Olson/Shutterstock

BEST PRACTICES BEFORE AN INSPECTION

Assign a leader

The inspector will ask to speak with the manager or person in charge. Staff should know who is in charge if the food manager is absent on the day of inspection. Management should always assign someone to take responsibility if the person in charge or the food manager is unavailable.

Have documentation

It's common for inspectors to request records concerning food purchases, pest control, or chemical use. Documentation provided during an inspection may become public. Before inspection, determine which documents are appropriate to provide and which documents are not. Review the company's policy regarding confidential information, or consult with an attorney when needed.

Prepare documentation.

Obtain a copy of the regulatory authority's inspection form and do a test run of a typical inspection. Staff can take turns playing the role of inspector. Review the results of the test run as a group, and make changes as needed before the real inspector arrives.

Self-audit

A daily inspection can be a powerful tool. Be sure that:

- The facility and equipment are clean and in good repair.
- Soap, paper towels, and cleaning materials are available.
- Staff follows hygiene rules and wears clean, protective clothing.
- Signage is satisfactory for the staff and the public.
- Food is in good condition and within its shelf life.
- Food and equipment temperatures are satisfactory.
- There is no potential for cross-contamination.
- No traces exist of pests or other hazards.
- All controls are in place, all records are complete, and staff follows procedures.

Conduct self-audits.

BEST PRACTICES DURING AN INSPECTION

Think of the inspection as a learning opportunity that will benefit the establishment and customers.

- Most inspectors volunteer their credentials immediately upon arrival, but criminals may try to gain entry into the establishment by posing as inspectors. Always ask for an inspector's credentials, and contact his or her supervisor for verification.

- It's completely reasonable to ask inspectors whether they're conducting a routine inspection or if they've come because of a customer complaint. Knowing why they've come will help to facilitate the inspection.

- Most inspectors can and will obtain an inspection warrant, which can't be refused. It doesn't make good sense to refuse entry to an inspector.

- Being defensive or uncooperative during an inspection may cause the inspector to think there is something to hide. Answer any and all questions the inspector asks, and require staff to do the same.

- Accompanying inspectors as they look around will show interest in what they have to say. It also provides the opportunity to correct minor violations on the spot.

- As the inspector makes suggestions or points out violations, write them down. While the inspector will provide a written report after the inspection, notes may be helpful for the future. Taking notes also shows the inspector that such comments will be taken seriously.

- If there is a disagreement with the inspector, don't start an argument. Keep quiet and appeal the decision later. The goal during the inspection is to make it clear that the facility will correct any violations.

- Avoid offering the inspector food or other items, which could be construed as an attempt to influence the inspection.

- Even the most reputable food facilities sometimes have violations. Don't be confrontational if there is a citation. Instead, ask the inspector to discuss the violations and the timeframe for correction. The discussion should involve both food management and staff.

- Signing the inspector's report does not mean there is an agreement about the findings. It means only that facility management received a copy of the report.

BEST PRACTICES AFTER AN INSPECTION

All violations outlined by the inspector must be corrected. Typically, there will be less than 48 hours to correct priority items. Correct any core and priority foundation items as quickly as possible.

Once the establishment is compliant, it's a best practice to determine why and how any violations occurred. Evaluate operating and training procedures to look for flaws. Revise existing procedures or implement new ones to eliminate problems.

SUSPENSIONS AND CLOSURES

If an inspector determines that the establishment poses an immediate and substantial threat to the public health, he or she can suspend the permit to operate or request a voluntary closure. In either instance, the establishment must immediately cease operation and eliminate hazards. Such hazards typically include significant sewer, water, or electrical issues; fire or flood concerns; insect or rodent infestation; or outbreaks of foodborne illness. Before resuming operation, the establishment must successfully pass one or more inspections.

The end result of a mandatory suspension or voluntary closure is the same—the establishment must cease operation. However, there are some important differences between the two actions.

Voluntary closures do not require official notice.

FeyginFoto/Shutterstock

An inspector can impose suspensions, but any suspension will require approval from the local health department. This means the establishment can request a hearing to dispute the suspension. A public notice is usually posted on the front door of an establishment when the permit to operate is suspended.

An inspector can usually request a voluntary closure without the approval of anyone else. The inspector is essentially asking the establishment to close until priority and priority foundation items are corrected. Voluntary closures do not typically require public notice.

Learning Objective: Identify best practices for sampling foods in a food facility.

The inspector may ask for a food sample to determine if any bacteria are present. Although the inspector will usually collect the sample, there may be instances when a manager is asked to provide one. The sample or specimen received for examination is very important. If samples are improperly collected or mishandled, or not representative of the sampled lot, the laboratory results will be meaningless. Because interpretations about a large consignment of food are based on a relatively small sample, sampling procedures must be applied uniformly. Obtaining a representative sample is essential when pathogens or toxins are sparsely distributed within the food. It is also important to have a representative sample to avoid having to unnecessarily dispose of an entire food shipment, when just one sample shows the number of bacteria found to be greater than the legal limit.

SAMPLING CONSIDERATIONS

The sample taken must be representative of the amount of product on hand. It is important to collect random portions from the same bag, box, or delivery to determine if contamination is widespread, or only from one food or type of food. The proper method of sampling—according to whether the food is solid, semisolid, viscous, or liquid—can be discussed with the regulatory authority.

Some things food managers should consider when food sampling is required include:

- Submit samples to the laboratory in original, unopened containers. If products are in bulk or in containers too large for submission, transfer representative portions to sterile containers under **aseptic** conditions.
- Use sterile sampling equipment and an aseptic technique.
- Use containers that are clean, dry, leakproof, wide mouthed, and sterile.
- Containers such as plastic jars or metal cans that are leakproof may be sealed. Whenever possible, avoid glass containers, which may break and contaminate the food product.
- For dry materials, use sterile metal boxes, cans, bags, or packets with suitable closures.

The sample taken must be representative of the amount of product on hand.
Alexander Raths/Shutterstock

- Sterile plastic bags (for dry, unfrozen materials only) or plastic bottles are useful containers for line samples. Do not overfill bags or puncture by wire closure.
- Identify each sample unit (defined later) with a properly marked strip of masking tape. Do not use a felt pen on plastic, because the ink can penetrate the container.
- Whenever possible, obtain at least 100 grams for each sample unit. Submit open and closed controls of sterile containers with the sample.

SAMPLE DELIVERY

Package the samples for delivery to maintain the original storage conditions as much as possible. When collecting liquid samples, take an additional sample as a temperature control. Check the temperature of the control sample at the time of collection, and again upon receipt at the laboratory. Make a record for all samples of the times and dates of collection. Dry or canned foods that are not perishable and are collected at ambient temperatures do not need to be refrigerated. Transport frozen or refrigerated products in approved, insulated containers of rigid construction so they will arrive at the laboratory unchanged. Collect frozen samples in pre-chilled containers.

Learning Objective: Explain the importance of using dates and labels on food.

Labeling includes any written, printed, or graphic matter appearing on food, on food containers or wrappers, or on any items accompanying the food.

LABELS

The ingredient list on a food label lists each ingredient in that food, in descending order of weight from highest to lowest weight. Any ingredient that consists of multiple ingredients must list each individual item within parentheses after the common name of the master ingredient. For example, the popular cookie ingredient might be listed as: chocolate chips (sugar, chocolate, cocoa butter, vanilla extract).

Food labels must indicate the food name and the use-by date. Whenever possible, food should be stored in its original packaging. If food is transferred to a new container, the container must be cleaned and sanitized. The container must be labeled with the food name and the original use-by date. The date must be clearly marked on the label to avoid serving or selling unsafe food. Labeling foods is especially important because certain foods can be difficult to identify if not in the original packaging. Inaccurate labeling can lead to allergic reactions when customers consume food items that are improperly labeled. It can even result in accidental chemical poisoning.

Federal law states that a food may be misbranded if the labeling is false or misleading in any particular way. A label does not have to specifically list a missing ingredient or leave off an included ingredient in order to be coined "misbranded." For example, under FDA and USDA policy, a food that claims to be natural but contains an artificial or synthetic ingredient may be considered misbranded. Additionally, an allergen, if recognized as one of the **Big Eight allergens**, must be listed as such. A food containing one of the Big Eight allergens that does not specifically disclose that ingredient as an allergen is considered misbranded. Undisclosed allergens on labels is the number-one reason food products are recalled.

| # HAZARDOUS MATERIALS

Learning Objective: Describe proper safety precautions for hazardous materials.

OSHA REQUIREMENTS

Some materials essential to a food establishment are hazardous, but only the hazardous materials required for a food service operation are allowed inside a food facility. Detergents, sanitizers, pesticides, and other chemicals can be harmful when not used properly. The Occupational Safety and Health Administration, or OSHA, is the federal agency that sets standards for the use of hazardous materials in the workplace. Some states have created OSHA-approved plans and set their own standards. Always ensure compliance by checking with a local regulatory agency.

Regardless of where the plan originated, food facilities must follow the Hazard Communication Standard, also known as HCS or HAZCOM. The HCS provides employees the right to know what hazardous materials are in their workplace and how such materials should be handled.

A vital part of the HCS is the availability of safety data sheets, or SDS (formerly called material safety data sheets, MSDS). These are created by chemical manufacturers, and they contain information about the hazards of specific chemicals and directions for safe use. OSHA requires that the SDS for each chemical be available to all employees in their work area. All employees should be trained on how to read safety data sheets.

OSHA requires that chemical manufacturers label their containers with:

- The physical, personal health, and environmental health hazards
- Protective measures
- Safety precautions for handling, storing, and transporting the chemical

The information contained in the SDS must be in English (although it may be in other languages as well). If the chemical is put into a different container, it must be labeled with the same information.

STORAGE AND EQUIPMENT

Chemicals and hazardous materials, including pesticides, cleaning materials, and sanitizers, must be in a restricted area or secured by a lock. These materials must be tracked using a daily inventory and usage log, and any discrepancies between the inventory log and what is found must be investigated immediately. Procedures must be in place to receive, securely store, and dispose of chemicals.

Personal protective equipment, commonly referred to as PPE, is equipment worn to minimize exposure to serious workplace injuries and illnesses. Injuries and illnesses may result from contact with chemical, physical, electrical, mechanical, or other workplace hazards. Personal protective equipment in a food facility may include items such as gloves, safety glasses, shoes, earplugs or muffs, and coveralls. All personal protective equipment should be safely designed and constructed, and it should be maintained in a clean and reliable fashion. It should fit comfortably, encouraging worker use. If the personal protective equipment does not fit properly, it can make the difference between being safely covered or dangerously exposed. Employers are required to train each worker when to use PPE and how to use it correctly.

FOOD MANAGER RESPONSIBILITIES

Responsibilities with respect to government regulations include:

- Knowing the agencies involved in the U.S. food safety system
- Determining which federal, state, and local agencies regulate the facility
- Obtaining copies of any regulations that the facility is subject to

Responsibilities concerning food safety regulations include:

- Knowing the agencies involved in the food safety system
- Determining which government agencies regulate the facility
- Obtaining copies of any regulations that the facility is subject to

Responsibilities concerning the state or federal food code include:

- Obtaining the most recent issue of the state food code that is based on the FDA Food Code
- Becoming familiar with all topics covered in the state food code

Responsibilities concerning HACCP programs include:

- Becoming familiar with HACCP principles
- Determining whether the facility is subject to regulations based on the HACCP principles
- Considering how the establishment can benefit from a HACCP program

Responsibilities concerning inspections include:

- Finding out how to access food safety data for the facility's jurisdiction
- Obtaining copies of the facility's previous inspection reports
- Determining how often the facility is subject to inspection
- Becoming familiar with local regulatory authority's inspection process
- Finding out what type of inspection the facility is subject to
- Training a staff member to act if the food manger is absent during an inspection
- Obtaining a copy of the local regulatory authority's inspection form
- Implementing a self-inspection system for the food facility

KEY TERMS

Aseptic   Free from microorganisms.

Core items   General physical facility conditions and general work practices that do not have a direct impact on food safety.

Food safety policy   Outlines management's responsibilities and communicates standards to staff.

Person in charge (PIC)   The individual present at a food establishment who is responsible for the operation at the time of inspection.

Personal protective equipment   Equipment worn to minimize exposure to serious workplace injuries and illnesses.

Priority foundation items   A provision whose application supports, facilitates, or enables priority items; a specific action or procedure to attain control of a hazard.

Priority items   An operation that contributes directly to the elimination, prevention, or reduction of hazards associated with foodborne illness or injury. These items include things such as cooking, reheating, cooling, and handwashing.

REFERENCES

1   Real progress in food code adoption. (2015, December 12). Retrieved from http://www.fda.gov/downloads/Food/GuidanceRegulation/RetailFoodProtection/FoodCode/UCM476819.pdf

ASSESSMENT QUESTIONS

1. The FDA Food Code is published every:
   a. Year
   b. Two years
   c. Four years
   d. Six years

2. How should information regarding a prepared food's ingredients and use-by date be communicated to staff?
   a. Memo
   b. E-mail
   c. Labels
   d. Posters

3. A traditional inspection is based on:
   a. An item system
   b. A point system
   c. A risk system
   d. Correct procedures

4. In a risk-based inspection, items that don't have a direct impact on food safety are called:
   a. Core items
   b. Violations
   c. Priority items
   d. Priority foundation items

5. SDSs are:
   a. Created by the manufacturer
   b. Required by OSHA
   c. Required for each chemical
   d. All of the above

6. In an inspection, what is the time limit to correct priority foundation items?
   a. 72 hours
   b. 24 hours
   c. Ten days
   d. Five days

7. HACCP-based inspections focus on the control of:
   a. Managers' interactions with waiters, chefs, and line cooks
   b. Hazards throughout the food flow
   c. Average wait times for tables
   d. TCS foods

8. The person in charge (PIC) is defined in the Food Code as:
   a. The individual present at a food establishment who is responsible for the operation at the time of inspection
   b. The individual at a food establishment with the most experience
   c. The individual present at a food establishment with the least uniform violations
   d. Another term for host or hostess

9. Obtaining a representative sample is essential when:
   a. The establishment serves more than 30 customers in an hour
   b. Pathogens or toxins are sparsely distributed within the food
   c. The PIC is new to the establishment
   d. The health inspector is expected to make a visit to the establishment

10. Personal protective equipment (PPE) is:
    a. Not necessary in most small food establishments
    b. Optional for experienced employees
    c. Equipment worn to minimize exposure to serious workplace injuries and illnesses
    d. An employee's clothes worn to and from the food establishment

QUESTIONS FOR DISCUSSION

1. Consider the volume of topics covered in the FDA Food Code. In your opinion, which topic requires the most attention? Why?

2. Imagine you are the PIC at a busy burger chain when the health inspector makes a visit. What are some best practices during the inspection?

3. Does cleaning and sanitizing a sampling container make it "sterile"? Why or why not? Explain the differences between clean, sanitized, disinfected, and sterile.

4. Federal law states that a food may be misbranded if the labeling is false or misleading in any particular way. List two or three examples of products that could be misbranded. What effect does misbranding have on the consumer?

5. What are some jobs that might require PPE? What is the appropriate PPE to use for each job?

CHAPTER ASSESSMENT ANSWER KEY

CHAPTER 1

1. D
2. A
3. B
4. A
5. A
6. D
7. A
8. B
9. B
10. D

CHAPTER 2

1. D
2. C
3. B
4. D
5. C
6. C
7. A
8. D
9. B
10. A

CHAPTER 3

1. B
2. B
3. D
4. A
5. B
6. C
7. D
8. B
9. C
10. C

CHAPTER 4

1. D
2. B
3. D
4. B
5. D
6. A
7. D
8. C
9. C
10. B

CHAPTER 5

1. C
2. C
3. A
4. D
5. D
6. B
7. B
8. A
9. B
10. C

CHAPTER 6

1. C
2. B
3. B
4. B
5. D
6. C
7. C
8. A
9. D
10. B

CHAPTER 7

1. C
2. D
3. A
4. D
5. D
6. A
7. A
8. C
9. C
10. D

CHAPTER 8

1. B
2. B
3. D
4. D
5. C
6. C
7. B
8. D
9. A
10. C

CHAPTER 9

1. A
2. B
3. C
4. A
5. C
6. D
7. B
8. D
9. B
10. D

CHAPTER 10

1. C
2. C
3. B
4. A
5. D
6. C
7. B
8. A
9. B
10. C

GLOSSARY

ACIDIC Having a pH level less than 7, as do foods such as vinegar and tomatoes.

ACTIVE MANAGERIAL CONTROL (AMC) The purposeful incorporation of specific actions or procedures by industry management into the operation of their businesses in order to attain control over foodborne illness risk factors.

ADA The Americans with Disabilities Act is a federal civil rights law that prohibits discrimination against people with disabilities.

ADULTERATED FOOD Food that is generally impure, un-safe, or unwholesome.

AEROBE An organism that requires oxygen to multiply. Also referred to as an aerobic organism.

AFLATOXINS A type of mycotoxin found in moldy peanuts, seeds, and spices that cannot be killed by cooking.

AIR GAP The vertical air space that separates the end of a supply line and the flood level rim of a sink, drain, or tub. It is one of the cheapest and most reliable methods of backflow prevention.

ALKALINE Having the opposite chemical property from acids. Alkaline products have a pH level greater than 7.

ALLERGEN Any substance that can cause an allergic reaction in some people, when their immune system sees the substance as foreign or dangerous.

AMBIENT TEMPERATURE The temperature of the surrounding environment.

ANAEROBE An organism that requires the absence of oxygen to multiply. Also referred to as an anaerobic organism.

ANAPHYLACTIC REACTION A severe allergic reaction affecting the whole body, often within minutes of eating the food, which may result in death. Also referred to as anaphylaxis.

ASEPTIC Free from microorganisms.

BACILLUS CEREUS (B. CEREUS) Intoxication-causing bacteria commonly found in starchy foods and meat products. This type of bacteria produces two types of toxins: emetic and diarrheal.

BACK PRESSURE Backflow due to a device, such as a boiler, that generates pressure greater than that of the drinking water supply.

BACK SIPHONAGE Backflow due to a sudden drop in the supply pressure of a water main, causing contaminated water to reverse-flow back into the water main.

BACKFLOW The reverse flow of water from a contaminated source to the drinking water supply. It can occur from back pressure or back siphonage.

BACTERIA Single-celled microorganisms with rigid cell walls that multiply by dividing in two; that is, by binary fission. Some bacteria cause illness, and others cause food spoilage.

BACTERIAL SPORES A resistant resting phase of bacteria, protecting them against adverse conditions such as high temperatures.

BIG EIGHT ALLERGENS The major food allergens: milk, eggs, fish, shellfish, tree nuts, peanuts, wheat, and soy. These foods account for about 90 percent of all food allergies in the United States.

BINARY FISSION The method by which bacteria multiply; they split into two.

BIOLOGICAL CONTAMINATION Food contamination by microorganisms, including bacteria, viruses, parasites, and fungi.

BLAST CHILLER Rapid cooling refrigeration unit.

CAMPYLOBACTER JEJUNI Infection-causing bacteria found on raw poultry and in contaminated water.

CARRIER A person who harbors, and may transmit, pathogenic organisms with or without showing any signs of illness.

CDC Centers for Disease Control and Prevention.

CHEMICAL CONTAMINATION The contamination of food by chemical substances such as pesticides and cleaning solutions. Includes contamination by natural toxins and allergens.

CIGUATOXIN A toxin found in some tropical coral reef fish. The toxin causes the following symptoms when consumed: nausea, vomiting, diarrhea, muscular weakness, numbness in extremities, and possibly respiratory arrest.

CIP Stands for cleaning in place. This cleaning process is necessary when equipment cannot be dismantled or moved. It involves the circulating of non-foaming detergents and disinfectants, or sanitizers, through assembled equipment and pipes, using heat and mostly turbulence to attain a satisfactory result.

CLEANING The process of removing soil, food residues, dirt, grease, and other objectionable matter; the chemical used to do this is called a detergent.

CLEANING AGENT A chemical compound, such as soap, that is used to remove dirt, food, stains, or other deposits from surfaces.

CLOSTRIDIUM BOTULINUM An anaerobic, intoxication-causing bacteria commonly found in soil and therefore in products that come from soil, such as root vegetables. It can also be found in improperly canned food.

CLOSTRIDIUM PERFRINGENS A bacteria that causes mild infection from toxin-producing spores. It is anaerobic and can be found in soil, animal and human waste, dust, insects, and raw meat.

COMMINUTED Reduced in size by grinding, mincing, chopping, or flaking. Ground meats are examples of comminuted food.

CONTAMINATION The presence of physical, chemical, or biological matter in or on food or in the food environment.

CONTROL MEASURES The actions required to prevent or eliminate a food safety hazard or reduce it to an acceptable level.

CORE ITEMS General physical facility conditions and general work practices that do not have a direct impact on food safety.

CORRECTIVE ACTION The action to be taken when a critical limit is breached.

CRITICAL CONTROL POINT (CCP) A point or procedure in a food system where control can be applied and is essential to prevent or eliminate a food safety hazard or reduce it to an acceptable level.

CRITICAL LIMIT The value of a monitored action that separates the acceptable from the unacceptable.

CROSS-CONNECTION The mixing of drinking and contaminated water in plumbing lines.

CROSS-CONTAMINATION Cross-contamination occurs when bacteria from contaminated foods (usually raw) transfers to other foods by direct contract, drip, or indirect contact.

CRYPTOSPORIDIUM PARVUM A parasite found in soil, food, or water, or on surfaces that have been contaminated with infected human or animal feces.

DECLINE PHASE The period in the life cycle of bacteria during which more bacteria are dying than multiplying, leading to an overall decrease in their number.

DECOMPOSITION The process of decay, or breaking down of organic matter.

DETERGENT A chemical or mixture of chemicals made of soap or synthetic substitutes. It facilitates the removal of grease and food particles from dishes and utensils and promotes cleanliness so that all surfaces are readily accessible to the action of sanitizers.

DISINFECTION The process of destroying microorganisms to a level in which all except for bacterial spores are killed; the chemical used is called a disinfectant.

EMETIC Causing vomiting.

ENDOTOXIN A toxin present in the cell wall of many bacteria that is released upon the death of the bacteria.

EPA Environmental Protection Agency.

ESCHERIA COLI (E. COLI) A bacteria found in the intestines of mammals. It can be found in ground beef and contaminated produce.

EXCLUSION Requiring a worker to leave the food establishment as a result of specific illnesses, symptoms, or exposure to certain diseases.

EXOTOXIN A toxin produced during the multiplication of some bacteria. They are highly toxic proteins and are often produced in food.

FACULTATIVE ANAEROBE An organism that can multiply with or without the presence of oxygen.

FAT TOM The acronym that represents the conditions that support the rapid growth of bacteria. These conditions are food, acidity, time, temperature, oxygen, and moisture.

FDA Food and Drug Administration.

FIFO The acronym used for "first in, first out," which is a system used in stock rotation.

FLOW OF FOOD The path and direction that food follows through a food facility.

FOOD ADDITIVES Preservatives, food colorings, and flavorings that are added to food.

FOOD ALLERGY An identifiable immunological response to food or food additives, which may involve the respiratory system, the gastrointestinal tract, the skin, or the central nervous system.

FOOD ESTABLISHMENT Any business whose commercial operations deal with food or food sources.

FOOD SAFETY A scientific discipline that describes the handling, preparation, and storage of food in ways that prevent foodborne illness.

FOOD SAFETY HAZARD A biological, chemical, or physical agent in food, or a condition of food, with the potential to cause harm (that is, an adverse health effect) to the consumer.

FOOD SAFETY MANAGEMENT SYSTEM The policies, procedures, practices, controls, and documentation that ensure that food sold by a food business is safe to eat and free from contaminants.

FOOD SAFETY MODERNIZATION ACT (FSMA) The act that aims to ensure the U.S. food supply is safe by shifting the focus from responding to contamination to preventing it.

FOOD SAFETY POLICY Outlines management's responsibilities and communicates standards to staff.

FOOD SPOILAGE BACTERIA Bacteria that diminish food quality, but rarely cause serious illness.

FOODBORNE ILLNESS An infection or intoxication that results from consuming foods contaminated with harmful microorganisms or toxins.

FOODBORNE ILLNESS OUTBREAK An instance when two or more cases of foodborne illness occur during a limited period of time with the same organism and are associated with either the same food service operation, such as a restaurant, or the same food product.

FUMIGATION A method of pest control that completely fills an area with smoke, gas, or vapor in order to kill vermin or insects.

FUNGI Biological contaminants that can be found naturally in air, plants, soil, and water. Fungi can be small, single-celled organisms or larger, multicellular organisms, and include molds and yeasts.

GENERATION TIME The time between bacterial divisions.

GERMINATE Also known as germination; the development or growth of microorganisms.

GIARDIA INTESTINALIS Also, *Giardia duodenalis*, *Giardia lamblia*; a parasite found in contaminated water, raw fruits, and vegetables.

HACCP Acronym for hazard analysis and critical control point. A food safety management system that identifies, evaluates, and controls hazards that are significant for food safety.

HACCP PLAN Documentation completed during the HACCP study and implementation. It includes the hazard analysis, the flow diagram, the HACCP control charts, monitoring records, verification details, and modifications to the system.

HAND WASHING The process of cleansing the hands with soap and water to thoroughly remove soil and/or microorganisms. Food workers must clean their hands up to their elbows.

HAZARD ANALYSIS The process of collecting and evaluating information on hazards and on conditions leading to their presence to decide which are significant for food safety and therefore should be addressed.

HEPATITIS A A virus primarily found in the feces of infected persons. It is spread from infected food workers to ready-to-eat food, including deli meats. It can also be spread to produce and salads and can be found in raw shellfish.

HIGH-RISK POPULATIONS People who are more likely to contract foodborne illness, including the elderly, the very young, people who are immunocompromised, and pregnant women.

HISTAMINE A naturally occurring substance produced in the body as an immune response to an allergen.

HIV A retrovirus spread through blood and bodily fluids. The CDC has found no evidence that HIV can be transmitted through food.

HOT HOLDING The storage of cooked food at 135°F (57°C) or higher while awaiting consumption by customers.

ICE BATH The method of cooling food in which a container holding hot food is placed into a sink or larger container of ice water. The ice water surrounding the hot food container disperses the heat quickly.

ICE PADDLES Plastic paddles filled with ice or water and then frozen; they are used to stir hot food to cool it quickly.

IMMUNOCOMPROMISED Having an immune system that is impaired, including the very old, the very young, and those with a disease or ongoing treatment that weakens the immune system.

INFECTION A disease caused by the release of endotoxins in the intestine of the affected person.

INFECTIOUS Communicable; tending to spread between people.

INTEGRATED PEST MANAGEMENT (IPM) An approach to pest control that uses a wide range of practices to prevent and solve pest problems in food facilities.

INTOXICATION An illness caused when bacteria produce exotoxins that are released into food.

JAUNDICE A yellowish discoloration of the skin and eyes, indicating liver malfunction and illness.

LAG PHASE The period in the life cycle of bacteria during which bacteria are not multiplying at all.

LESION A skin injury, usually caused by disease or trauma.

LISTERIA MONOCYTOGENES Infection-causing bacteria naturally found in soil, raw vegetables, and milk that has not been properly pasteurized. It is associated with certain ready-to-eat foods, such as deli meats and hot dogs.

LOGARITHMIC PHASE Also called log phase; the period during bacterial growth in which bacteria multiply rapidly.

MAP Acronym for modified atmosphere packaging. A physical preservation method that involves changing the proportion of gases normally present around a food item.

MICROBIOLOGICAL Of the branch of biology dealing with the structure, function, uses, and modes of the existence of microorganisms.

MICROORGANISMS Organisms, or living things, such as bacteria, viruses, fungi, and parasites that are too small to be seen with the naked eye. These microorganisms may contaminate food and cause foodborne illness.

MINIMUM INTERNAL TEMPERATURE The required minimum temperature that the internal portion of food must reach to sufficiently reduce the number of microorganisms that might be present. This temperature is specific to the type of food being cooked.

MOLD Microscopic fungi that produce threadlike filaments; mold can be black, white, or of various colors.

MONITORING The planned observation and measurement of control parameters to confirm that processes are under control and that critical limits are not exceeded.

MYCOTOXINS Poisonous substances produced by certain fungi.

NOAA National Oceanic and Atmospheric Administration (NOAA Fisheries is also known as the National Marine Fisheries Service, or NMFS).

NONCONTINUOUS COOKING Occurs when the initial heating of food is intentionally stopped, cooled, or held prior to sale or service.

NOROVIRUS Found in the feces of infected persons. Can also be found in contaminated water. Norovirus is the most common cause of viral gastroenteritis in humans.

ONSET TIME The period between eating contaminated food and the first signs of illness.

PARASITE An organism that lives and feeds in or on another living creature, known as a host, in a way that benefits the parasite and disadvantages the host.

PASTEURIZATION The process of applying heat to destroy pathogens.

PATHOGEN Disease-producing organism.

PERISHABLE FOODS Those foods subject to spoilage or decay.

PERSON IN CHARGE (PIC) The individual present at a food establishment who is responsible for the operation at the time of inspection.

PERSONAL HYGIENE Standards of personal cleanliness habits, including keeping hands, hair, and body clean and wearing clean clothing in the food establishment.

PERSONAL PROTECTIVE EQUIPMENT Equipment worn to minimize exposure to serious workplace injuries and illnesses.

PEST An animal, bird, or insect capable of directly or indirectly contaminating food.

PEST CONTROL OPERATORS (PCO) Individuals licensed to control pests in the state in which they operate.

PESTICIDES Chemicals used to kill pests.

PH An index used as a measure of acidity/alkalinity, measured on a scale of 1 to 14. Acidic foods have pH values below 7 and alkaline foods have values above 7; a pH value of 7 is neutral.

PHYSICAL CONTAMINATION When any foreign object is in or on a food and presents a hazard or nuisance to those consuming it.

POTABLE WATER Water that is safe to drink; an approved water supply.

PREREQUISITE PROGRAMS Procedures, including standard operating procedures (SOPs), that address basic operational and sanitation conditions in an establishment.

PRIONS Small pathogenic proteins that, like viruses, require a living host to multiply.

PRIORITY FOUNDATION ITEMS A provision whose application supports, facilitates, or enables priority items; a specific action or procedure to attain control of a hazard.

PRIORITY ITEMS An operation that contributes directly to the elimination, prevention, or reduction of hazards associated with foodborne illness or injury. These items include things such as cooking, reheating, cooling, and handwashing.

QUATERNARY AMMONIUM COMPOUNDS A common example of chemical disinfectants; also referred to as quats.

READY-TO-EAT (RTE) FOOD Food that is going directly to the consumer without further cooking or preparation to kill potential dangerous microorganisms.

REHEATING The process of heating previously cooked and cooled foods to a temperature of at least 165 °F (74 °C).

RESTRICTION Preventing a worker with certain illnesses or symptoms from working with food or in food contact areas.

RISK The likelihood of a hazard occurring.

RISK ASSESSMENT The process of identifying hazards, assessing risks and severity, and evaluating their significance.

ROP Refers to any packaging procedure that results in a reduced oxygen level.

ROUTE OF CONTAMINATION The path along which contaminants are transferred from their sources to food.

SAFE FOOD Food that is free of contaminants and does not cause harm to the person consuming it.

SALMONELLA, SPP. Several species of infection-causing bacteria commonly found in raw poultry, eggs, raw meat and dairy products. It has also been found in ready-to-eat food that has come into contact with infected animals or their waste.

SANITIZATION Use of chemicals or heat to reduce the number of microorganisms to a safe level.

SCOMBROTOXIN A toxin that forms when certain fish aren't properly refrigerated before being processed or cooked. Examples of fish that can form the toxin if they start to spoil include tuna, mahi mahi, bluefish, sardines, mackerel, amberjack, and anchovies.

SHIGA TOXIN–PRODUCING E. COLI An infection-causing bacteria, containing the strain 0157:H7, also known as STEC, found in ground meats, unpasteurized raw milk, and contaminated produce.

SHIGELLA, SPP. Several species of bacteria found in the feces of people with Shigellosis. It can be found in ready-to-eat foods such as greens, milk products and vegetables and also in contaminated water. The most common method of transmission is cross contamination. Flies can also be carriers of this type of bacteria.

SLACKING The process of gradually increasing the temperature of frozen food from −10°F to 25°F (−23°C to −4°C); generally in preparation for deep frying.

SPOILAGE ORGANISM An organism that damages the nutrition, texture, and flavor of the food, making it unsuitable to eat.

SPORE A resistant resting phase of bacteria, protecting them against adverse conditions such as high temperatures.

STAPHYLOCOCCUS AUREUS An intoxication-causing bacteria commonly found on the skin, nose and hands of one out of two people. It is transferred easily from humans to food when people carrying the bacteria handle the food without washing their hands. This bacteria also produces toxins that multiply rapidly in room-temperature food.

STATIONARY PHASE The period in the life cycle of bacteria during which the number of bacteria produced is equal to the number of bacteria dying.

STOCK ROTATION The practice of ensuring the oldest stock is used first and that all stock is used within its shelf life.

TARGET LEVEL The predetermined value for the control measure that will eliminate or control the hazard at a control point.

TEMPERATURE DANGER ZONE The temperature range at which most foodborne microorganisms rapidly grow. The temperature danger zone is 41°F to 135°F (5°C to 57°C).

TIME/TEMPERATURE CONTROL FOR SAFETY FOODS (TCS) Products that under the right circumstances support the growth of microorganisms that cause foodborne illness.

TOXIC Directly poisonous; affected by a toxin or poison.

TOXIC METAL POISONING The leaching of certain poisonous metals, such as aluminum or copper, into acidic foods being prepared with pots and/or utensils of those metals.

TOXIN A substance created by plants or animals that is poisonous to humans.

TOXOPLASMA GONDII Intracellular parasites. According to the CDC, this parasite is the second leading cause of death from foodborne illness in the U.S. The illness it causes, toxoplasmosis, can be serious or deadly, particularly for babies infected in the womb and people with weak immune systems.

TRANSMISSION The process of spreading (as in a disease or infection) from person to person.

TRICHINELLA SPIRALIS An intestinal roundworm that is found in wild game animals and in undercooked pork. The larvae of the *Trichinella spiralis* can move throughout the body, infecting various muscles and causing the infection trichinosis.

TRICHINOSIS An infection caused by *Trichinella spiralis*.

USDA United States Department of Agriculture.

USE-BY DATE This is the last date recommended for the use of the product if it is to be used at peak quality. The manufacturer of the product usually determines the date. A retail facility may not sell any food past the use-by date.

VALIDATION The element of verification focused on collecting and evaluating scientific and technical information to determine if the process or procedure will effectively control the hazards.

VEGETATIVE STATE The condition in which bacteria divide at regular intervals due to surroundings suitable for their growth and multiplicationn.

VERIFICATION The application of methods, procedures, and tests to determine compliance with a food safety plan, in addition to those used in monitoring to determine compliance with a HACCP plan.

VIBRIO PARAHAEMOLYTICUS An infection causing bacteria commonly associated with raw or partially cooked oysters.

VIRAL GASTROENTERITIS The swelling or inflammation of the stomach and intestines from a virus, leading to diarrhea and vomiting.

VIRUSES Submicroscopic pathogens that multiply in the living cells of their host.

WAREWASHING The washing of the wares, or in the food industry, food contact materials.

WATER ACTIVITY A measure of the moisture in food available to microorganisms. It is represented by the symbol a_w. Most bacteria multiply best in food with a water activity of between 0.95 and 0.99.

YEASTS Single-celled microscopic fungi that reproduce by budding and that grow rapidly on certain foodstuffs, especially those containing sugar.

INDEX

A

Abrasive cleaners, 87
Accessibility guidelines, 81
Acid cleaners, 86
Acidic, defined, 42
Acidic foods:
    bacterial growth and, 26
    toxic metal poisoning from, 35
Active managerial control (AMC),
130, 138
ADA (Americans with Disabilities Act),
81, 97
Additives, 5, 6, 117
Adulterated food, 2, 6
Aerobes, 26, 42
Aflatoxins, 34, 39, 42
Air curtain, 48
Air gaps (plumbing), 96, 97
Alarms, refrigerator, 105
Alkaline, defined, 42
Alkaline foods, 26
Allergens, 40–41, 42, 150
Allergies:
    food, 40–41
    latex, 68
Aluminum containers, 35
Ambient temperature, 102, 110
AMC (Active managerial control),
130, 138
American National Standards Institute
(ANSI), 84
Americans with Disabilities Act (ADA),
81, 97
Amnesic shellfish poisoning (ASP), 39
Anaerobes, 26, 42
Anaphylactic reaction, 40, 42
Animals, 47, 48
Anisakis simplex, 33
ANSI (American National Standards
Institute), 84
Ants, 48
Aprons, 62
Aseptic (term), 148, 152
ASP (amnesic shellfish poisoning), 39
Association of Food and Drug
Officials, 142
Asthma, 41
Audits:
    cooling procedures, 123
    daily pre-inspection, 145
    dry storage areas, 107
    food preparation, 117
    freezers, 107
    in HACCP plan, 135, 137

    prepared food stocks, 126
    refrigerators, 106
    reheating procedures, 124
    service, 126
Avocados, storage of, 110

B

Bacillus cereus, 23, 24, 27, 42
Backflow, 96, 97
Back pressure, 96, 97
Back siphonage, 96, 97
Bacteria:
    carriers and sources of, 24
    defined, 23, 42
    growth phases and variables,
    25–27
    spores, 24, 42, 43
    structure of, 24
    types of, 23, 27–28
Bacterial growth
    as delivery hazard, 102
    during food service, 125
    freezing effect on, 106
    from time/temperature abuse, 114
Bacterial illness classification, 25
Baker's yeast, 32
Bananas, storage of, 110
Bare-hand contact, 67
Batch cooking, 117
Beans, toxins in, 37
Beef:
    cooking, 121, 122
    holding roast beef, 125
    storage of, 105, 108
Big Eight allergens, 40–41, 42, 150
Bi-metal stem thermometers,
115–116
Binary fission, 25, 42
Biological contamination:
    bacterial, 23–28
    defined, 18, 42
    fungal, 32–34
    parasitic, 32, 33
    prionic, 31
    viral, 29–31
Birds, 47, 48
Blast chillers, 123, 126
Botulism, 10
bovine spongiform encephalopathy
(mad cow disease), 31
Brainstorming, in hazard analysis, 132
Break areas, 52
Buildings. *See* Facilities
Butter, frozen storage of, 106

C

Campylobacter jejuni, 23, 27, 42
Canned food, 102, 107
CAP (Controlled atmosphere
packaging), 110
Capsule (bacteria), 24
Carbonated beverage systems, 36
Carpet, 78
Carriers, 24, 42
Cast iron, 83
CCPs (critical control points), 131,
134, 135, 138
CDC (Centers for Disease Control),
5, 6, 142
Ceilings, 53, 79
Cell wall (bacteria), 24
Centers for Disease Control (CDC),
5, 6, 142
Certified Food Protection Manager
(CFPM), 130, 142
CFP (Conference for Food Protection),
142
Cheese:
    frozen storage of, 106
    mold on, 34
Chemical cleaning, 86
Chemical contamination, 18–19,
35–36, 42, 78
Chemicals, use and storage of, 35
Chemical sanitization, 88–89, 92
Chicken. *See* Poultry
Children, and foodborne illness, 12
Chlorine, 88–89
Ciguatoxin, 37–38, 39, 42
CIP (cleaning in place), 94, 97
Citrus fruits, 27, 109
"Clean as you go" policy, 89
Cleaning:
    chemical contamination and, 36
    defined, 97
    methods of, 86
    for pest prevention, 52–53
    purpose of, 86
    of service equipment, 126
    written cleaning schedule, 89–90
Cleaning agents, 86–87, 97, 157
Cleaning in place (CIP), 94, 97
Cleaning tools and equipment, 53,
87, 88
Clostridium botulinum:
    in canned or ROP foods, 102, 110
    characteristics of, 23, 24, 27
    defined, 110
Clostridium perfringens, 23, 24, 27

Closures, post-inspection, 146–147
Clothing, protective, 62
Cocci, 24
Cockroaches, 47
Cold holding, 120, 125
Color-coding systems, 18, 85, 87, 117
Comminuted meats, 121, 126
Communication, 72
Condiments, reserving, 125
Conference for Food Protection (CFP), 142
Contact time, for sanitizers, 88
Contamination, 17–43
    allergens, 40–41
    biological, 18, 23–34
    checking deliveries for, 102–103
    chemical, 18–19, 35–36, 56
    defined, 18, 42
    employee hygiene and, 63–64
    facility design and, 78
    during food service, 125
    intentional, 20–22
    natural toxins, 37–39
    physical, 19–20
    during refrigerated storage, 104–106
Controlled atmosphere packaging (CAP), 110
Control measures, 132, 133, 138
Cooking:
    minimum internal temperatures, 121–122
    noncontinuous, 122, 126
    safe practices, 120
    thawing food during, 119
Cookware, 35, 120
Cooling foods, 123
Copper containers, 35
Core items, 144, 152
Corrective actions (HACCP), 131, 136, 138
Critical control points (CCPs), 131, 134, 135, 138
Critical limits, 131, 134, 138
Cross-connection (plumbing), 96, 97
Cross-contamination:
    defined, 18, 42
    facility design and, 78
    prevention strategies, 18, 85, 117
    from raw foods, 24, 78
    refrigerator organization and, 105, 106
    during thawing, 117
    transfer methods, 18
Cryptosporidium parvum, 32, 33, 42
Custards, bacterial growth and, 27
Customers:
    complaints and claims of, 137
    as contamination risk, 20
Cuts (wounds), 63
Cyclospora cayetanensis, 33

D

Daily inspections, 145
Dairy products. *See* Milk and dairy products
Date labels:
    checking during deliveries, 102
    food storage and, 104, 106, 107
    on prepared foods, 126
    for ROP foods, 110
Decline phase (bacterial growth), 25, 42
Decomposition, 32, 42
Defrosting of freezers, 107
Degreasers, 86
Delimers, 86
Delivery:
    hazards of, 102
    pest prevention and, 54
    physical contamination prevention, 19
    receiving shipments, 102–103
    shortcomings and rejections, 103
    temperature control during, 102–103
Delivery drivers, 101
Delivery notes, 103
Detergents, 86, 97
Diarrhetic shellfish poisoning (DSP), 39
Direct cross-contamination, 18
Dishwashers, 54, 92
Dishwashing procedures, 93–94
Dishwashing stations, 78, 92
Disinfection, 87, 97
Display cabinets, 20
Documentation and records:
    for bare-hand contact, 67
    of corrective actions, 136
    of employee training, 74
    HACCP, 136, 137–138
    for inspections, 145
    stock records, 104
    of suppliers, 101
Doors, and pest prevention, 53–54
Double-wash procedure (hands), 65
Drinking water, 95–96
Drip cross-contamination, 18
Dry storage, 107
DSP (diarrhetic shellfish poisoning), 39
Duck. *See* Poultry
Ductwork, 54
Dumpsters. *See* Garbage storage areas

E

Ears, hygiene and, 64
E. coli:
    characteristics of, 23, 27
    defined, 42
    employee restrictions and, 69
    Shiga toxin-producing, 69, 74
Eggs:
    cooking, 121
    as food allergen, 40
    purchasing, 101
    storage of, 108–109
    temperature control during delivery, 102, 103
Emergency guidelines (potable water supply), 95–96
Emetic toxin, 23, 42
Employee areas, 52
Employee screening, 21
Employee training, 61–74
    bare-hand contact, 67
    communication, 72
    contamination prevention, 19, 22
    documentation of, 74
    glove use, 68
    hand-washing, 65–66
    illness reporting, 69–71
    methods and value of, 73–74
    personal hygiene, 62–64
    refresher training, 74
Endotoxins, 25, 42
Environmental Protection Agency (EPA), 6
EPA (Environmental Protection Agency), 6
Equipment, 77–97
    for cooling foods, 123
    cross-contamination and, 18, 85
    dishwashers, 54, 86, 92
    for food service, 126
    maintenance and use of, 84–85
    non-food contact, 83
    purchasing, 84
    standards for, 82
    toxic metal poisoning from, 35
    utensils, 82–83
Exclusion, 69–71, 74
Exhaust hoods, 80
Exotoxins, 25, 42

F

Facilities, 77–97
    design and layout of, 78–81
    intentional contamination prevention in, 22
    pest prevention in, 53–54
    plumbing systems, 96–97
    washing facilities, 91–94, 117
    water supplies, 95–96
Facultative anaerobes, 26, 42
FALCPA (Food Allergen Labeling and Consumer Protection Act), 40, 41
Fans (exhaust hoods), 80
FAT TOM, 26, 42
FDA (Food and Drug Administration), 4–5, 6, 129, 142
FDA Food Code:
    bare-hand contact, 67
    carpeting, 78
    comminuted meats, 121
    development and updates of, 142
    employee illness flow charts, 70–71
    food manager responsibilities, 152
    HACCP principles in, 129
    lighting levels, 79
    screen specifications, 54
    state adoption of, 4

Federal government agencies, 4–6
Feedback, during training, 73
FIFO (first in, first out), 104, 110
Fingernails, 62
First aid equipment, 91
Fish. *See also* Shellfish
　cooking, 121, 122
　as food allergen, 40
　purchasing, 101
　scombrotoxic fish poisoning, 38, 39, 43
　smoked, 109
　storage of, 105, 106, 109
　temperature control during delivery, 103
　toxins in, 37–38
Flagella, 24
Flies, 48
Floor cleaning equipment, 53
Floor drains, 54
Floors and flooring:
　pest prevention and, 53
　types of, 78–79
Flow charts:
　employee illness reporting, 70–71
　HACCP, 132, 133
Flow of food, 100, 110
Food, for bacterial growth, 26
Food additives, 5, 6, 117
Food Allergen Labeling and Consumer Protection Act (FALCPA), 40, 41
Food allergies, 40–41, 42
Food and Drug Administration (FDA), 4–5, 6, 129, 142. *See also* FDA Food Code
Foodborne illness, 9–14
　bacterial, 25
　causes and symptoms of, 10
　defined, 10, 14
　high-risk populations, 12–14
　vs. outbreaks, 10
　prevention of, 3
　publicized cases, 1
　risk factors, 11
　statistics, 3
　viral, 29–31
Foodborne illness outbreak, 10, 11, 14
Food code. *See* FDA Food Code
Food contact materials, 82–83
Food contact surfaces, 88
Food defense program, 20–21
Food displays, 20
Food establishments, 11, 14
Food handling, 113–126
　bare-hand contact, 67
　cooking, 120–122
　cooling and reheating, 123–124
　employee health and, 69–71
　food preparation, 117–119
　gloves for, 68, 117
　service, 125–126
　time and temperature in, 114–116
Food labels. *See* Labeling
Food poisoning. *See* Foodborne illness

Food preparation:
　audits of, 117
　physical contamination risks, 19
　in process approach to HAACP, 131
　safe practices, 117
　thawing methods, 117–119
Food preparation areas:
　cross-contamination control in, 18
　pest control in, 52, 53
Food safety, 1–6
　benefits of, 3
　defined, 2, 6
　regulatory authorities, 4–6
Food Safety and Inspections Service (FSIS), 5
Food Safety and Modernization Act (FSMA), 3, 6
Food safety hazard, 63, 74
Food safety management system:
　defined, 110
　of suppliers, 101
Food safety policies:
　defined, 141, 152
　personal hygiene standards, 63, 69
　to prevent contamination, 19
　procedures included in, 141
Food safety standards. *See* Standards
Food serving areas, 53, 54
Food spoilage bacteria, 23, 42
Food storage, 104–110
　after delivery, 103
　chemical contamination and, 36
　cross-contamination and, 18
　dry storage, 107
　eggs, 108–109
　frozen storage, 106–107
　fruits and vegetables, 109–110
　labeling, 104, 106, 107
　meats, 108
　milk and dairy products, 109
　physical contamination and, 19
　poultry, 109
　reduced oxygen packaging foods, 110
　refrigerated storage, 104–106
　RTE and TCS foods, 126
　seafood, 109
　stock records, 104
　stock rotation, 104, 107, 110
Footwear, 62
Foreign objects, 19
Foster Farms, 9
Freezers, 106–107
Frozen food:
　bacterial growth in, 114
　storage of, 106–107
　temperature control during delivery, 103
　thawing methods for, 117–119
Fruits and vegetables:
　bacterial growth and, 27
　cooking, 121

　purchasing, 101
　storage of, 105, 106, 109–110
FSIS (Food Safety and Inspections Service), 5
FSMA (Food Safety and Modernization Act), 3, 6
Fumigation, 56, 58
Fungi, 32–34, 42

G

Game:
　commercially raised, 121
　wild, 32, 108
Garbage storage areas:
　flooring for, 78
　guidelines for, 80–81
　pest prevention in, 52, 55
Garnishes, re-serving, 125
Gastroenteritis, 17, 29, 43
Generation time (bacterial growth), 26, 42
Germination, of bacteria, 24, 42
Giardia duodenalis, 33
Giardia intestinalis, 32, 42
Giardia lamblia, 33
Gloves, 68, 117
Goose. *See* Poultry
Government food safety authorities, 4–6
Grease traps, 97
Ground meats, 105, 108, 121

H

HACCP (Hazard Analysis and Critical Control Point), 129–138
　active management controls in, 130
　corrective actions, 136
　critical control points, 134
　critical limits, 134
　defined, 129, 138
　documentation and record keeping, 137–138
　food manager responsibilities for, 152
　hazard analysis, 131, 132–133
　history of, 129
　monitoring system, 134–136
　overview, 130–131
　prerequisite programs, 130, 138
　verification procedures, 136–137
HACCP-based inspections, 144
HACCP documentation (of suppliers), 101
HACCP plan, 131, 138
HACCP team, 132
Hair, hygiene and, 64
Hand contact surfaces, 88
Hands. *See also* Hand washing
　bare-hand contact, 67
　hygiene and, 64
Hands-on activities, 73
Hand washing, 65–66, 74
Hand washing stations, 66, 91

Hantavirus, 47
Hard-surface flooring, 78
Hazard analysis, 131, 132–133, 138
Hazard Analysis and Critical Control
Point. *See* HACCP
Hazard Communication Standard
(HCS, HAZCOM), 151
Hazardous materials:
    cleaning agents, 86–87
    pesticides, 56–58
    standards for, 151
Health, of employees, 69–71
Health inspections. *See* Inspections
Heating, ventilation, and air
conditioning (HVAC) systems, 54, 80
Heat sanitization, 88
Hepatitis A virus, 30, 42, 69
High-risk populations, 12–14, 67, 69
High-temperature dishwashers, 92
Histamines, 40, 42
HIV virus, 31, 42
Holding foods, 120, 125
Hoods, 80
Hot holding, 120, 125, 126
Hot water heaters, 54
HSP (Highly-susceptible populations),
12–14, 67, 69
HVAC systems, 54, 80
Hygiene. *See* Personal hygiene

I

Ice, 125
Ice bath, 123, 126
Ice paddles, 123, 126
Illness. *See also* Foodborne illness
    caused by pests, 47
    in employees, 69–71
Immunocompromised, 12, 14
Indirect cross-contamination, 18
Infections:
    bacterial, 25
    defined, 42
    viral, 29–31
Infectious (term), 25, 42
Infrared thermometers, 116
Insects, 47, 48
Inspections. *See also* Audits
    best practices after, 146
    best practices before, 145
    best practices during, 146
    food manager responsibilities,
    152
    frequency and notice of, 143–144
    HACCP records and, 137
    purpose of, 143
    suspensions and closures from,
    146–147
    types of, 144
Integrated pest management (IPM),
49–51, 58
Intentional contamination, 20–22
Intoxication (bacterial), 25, 42

Iodine, 88–89
IPM (integrated pest management),
49–51, 58
IPM (integrated pest management)
team, 50

J

Jaundice, 30, 42
Jewelry, 62

L

Labeling:
    allergen requirements, 41
    for food storage, 106, 107
    of hazardous materials, 35, 151
    for ready-to-eat and TCS foods,
    126
    on received food deliveries, 102
    of ROP foods, 110
    standards for, 150
Lag phase (bacterial growth), 25, 42
Lamb, 108, 121
Landscape maintenance, 54
Language barriers, 72
Laser thermometers, 116
Latex gloves, 68
Layout, facility, 78–81
Leadership, 72
Leaky pipes, 97
Leftovers, cooking, 121
Leptospirosis, 47
Lesions, 63, 74
Lighting, 79
Listening, 72
Listeria monocytogenes:
    characteristics of, 23, 28
    defined, 42
    in ROP foods, 110
Local regulations, 4
Locker rooms, 52
Logarithmic phase (bacterial growth),
25, 42
Low-temperature dishwashers, 92
Lubricants, 36
Lyme disease, 47

M

Mad cow disease, 31
Maintenance practices, 20
Make-up air, 80
Manager responsibilities, 152
Manual dishwashing, 93–94
MAP (modified atmosphere
packaging), 110
Material safety data sheets (MSDSs),
56, 57, 151
Measurements (monitoring), 135
Meats:
    bacterial growth and, 27
    cooking, 121

storage of, 105, 106, 108
    temperature control during
    delivery, 102
Melons, bacterial growth and, 27
Menu advisories, 121
Mice, 47
Microbiological (term), 73, 74
Microorganisms, 18, 42. *See also*
Biological contamination
Microwave cooking, 120
Microwave method (thawing), 119
Milk and dairy products:
    bacterial growth and, 27
    as food allergen, 40
    purchasing, 101
    storage of, 109
Minimum internal temperature, 114,
121–122, 126
Misbranding, 2, 150
Modified atmosphere packaging
(MAP), 110
Moisture, and bacterial growth, 26, 27
Molds, 32–34, 42
Monitoring. *See also* Audits
    defined, 134, 138
    in HACCP plan, 131
    methods of, 135–136
Monosodium glutamate (MSG), 41
Mouth, hygiene and, 64
MSDSs (material safety data sheets),
56, 57, 151
MSG (monosodium glutamate), 41
Mushrooms, toxic, 37, 39
Mycotoxins, 34, 39, 42

N

NASA, 129
National Oceanic and Atmospheric
Administration (NOAA), 5, 6
Natural toxins, 37–39
Neurotoxin shellfish poisoning
(NSP), 39
Nightshade, 37
NOAA (National Oceanic and
Atmospheric Administration), 5, 6
Noncontinuous cooking, 122, 126
Non-drinking water, 95
Non-food contact equipment, 83
Nonporous, resilient flooring, 78
Nonslip flooring, 79
Nontyphoidal *Salmonella*. *See*
Salmonella spp.
Norovirus:
    defined, 42
    described, 29–30
    employee restrictions and, 69
    as gastroenteritis cause, 17
Nose, hygiene and, 64
NSF International, 84
NSP (neurotoxin shellfish
poisoning), 39
Nuts, as food allergen, 40

O

Observations, 135
Occupational Safety & Health Administration (OSHA), 151
Older adults, and foodborne illness, 12
Olives, and bacterial growth, 27
Onset time, 25, 42
OSHA (Occupational Safety & Health Administration), 151
Out-of-date foods, 104, 106, 107. *See also* Date labels
Outside dining areas, 54
Oxygen, for bacterial growth, 26

P

Paralytic shellfish poisoning (PSP), 39
Parasites, 32, 33, 42
Pasteurization, 101, 110
Pathogens, 23, 42
Patulin, 39
PCOs (pest control operators), 50–51, 58
Peanuts, as food allergen, 40
Perfume, 62
Perishable foods, 104, 110
Permit suspensions, 146–147
Personal hygiene:
    contamination prevention and, 63–64
    defined, 74
    hand washing, 65–66
    policies and standards for, 63, 69
    smoking, 62–63
    work attire, 62
Personal protective equipment (PPE), 151, 152
Person in charge (PIC), 143, 152
Pest control operators (PCOs), 50–51, 58
Pesticides:
    defined, 56, 58
    forms of, 56
    inventory and usage logs, 151
    selection and use, 56–57
    storage and disposal, 56, 58
Pests and pest control, 46–58
    common pests, 47–48
    definitions, 58
    in dry storage areas, 107
    integrated pest management, 49–51
    IPM team, 50
    lighting and, 79
    in outside dining areas, 54–55
    plumbing and, 97
    prevention strategies, 52–55
pH, 26, 43
Physical cleaning, 86
Physical contamination, 19–20, 43
PIC (person in charge), 143, 152

Pillsbury Company, 129
Pipes, leaky, 97
Plants, poisonous, 37
Plumbing system, 54, 96–97
Policies. *See* Food safety policies
Poliomyelitis, 47
Pork:
    cooking, 121, 122
    holding roast pork, 125
    storage of, 105, 106, 108
    undercooking risk, 32
Positive airflow, 80
Potable water, 95–96, 97
Potatoes, storage of, 107
Poultry:
    cooking, 121
    purchasing, 101, 108
    storage of, 105, 106, 109
    temperature control during delivery, 102
PPE (Personal protective equipment), 151
Pregnancy, and foodborne illness, 12
Pre-packaged foods, 126
Prep areas:
    cross-contamination control in, 18
    pest control in, 52, 53
Prerequisite programs, 130, 138
Prevention, vs. reaction, 3, 131
Prions, 31, 43
Priority foundation items, 144, 152
Priority items, 144, 152
Private water supply systems, 95
Process approach (HAACP), 131
Produce. *See* Fruits and vegetables
Product recalls, 150
Protective clothing, 62
Protective equipment, 151, 152
Protein channel, 29
Proteins (prions), 31, 43
PSP (paralytic shellfish poisoning), 39
Purchase orders, 103
Purchasing:
    equipment, 84
    food, 101

Q

Quality standards, for delivered foods, 103
Quaternary ammonium compounds (quats), 88–89, 97

R

Rats, 47
Raw foods:
    cross-contamination control, 18, 24, 78, 105, 117
    separating from cooling food, 123
    service of, 125
    storage of, 105

Ready-to-eat (RTE) foods:
    bare-hand contact with, 67
    cooking, 121
    cross-contamination control, 18, 24, 78, 105, 117
    defined, 18, 43
    labeling, 126
    storage and shelf life, 105
Recordkeeping. *See* Documentation and records
Reduced oxygen packaging (ROP) foods, 110
Reflective listening, 72
Refresher training, 74
Refrigerated foods:
    bacterial growth in, 114
    storage of, 104–106
    temperature during delivery, 102
Refrigerator method (thawing), 118
Refrigerators:
    audits of, 106
    cleaning, 105
    food placement in, 105–106
    hot foods in, 123
    walk-in, 78
Refuse disposal, 52. *See also* Garbage storage areas
Regulatory authorities, 4–6
Reheating, 124, 126
Re-serving of food, 125
Restriction, 69–71, 74
Restrooms, 53, 78, 91
Review stage (HACCP verification), 136–137
Rhubarb leaves, 37
Risk, 133, 138
Risk assessment, 133, 138
Risk-based inspections, 144
Roasts, 122, 125
Rodents, 47
Rods (bacteria), 24
Room temperature, 117
Root vegetables, storage of, 109
ROP (reduced oxygen packaging) foods, 102, 110
Route of contamination, 29, 43
RTE foods. *See* Ready-to-eat (RTE) foods
Rubber and rubber-like utensils, 82
Rubber floor mats, 79

S

Saccharomyces cerevisiae, 32
Safe food, 2, 6
Safety Data Sheet (SDS), 56, 57, 151
Salmonella spp.:
    carriers of, 47
    characteristics of, 23, 28
    employee restrictions and, 69
    publicized cases, 1, 9
Salmonella Typhi, 28, 69, 70
Sampling, 148–149

Sanitation Standard Operating Procedures (SSOPs), 90
Sanitization:
    for chemical contamination prevention, 36
    defined, 97
    equipment requiring, 88
    methods of, 88–89
    purpose of, 87
Sanitizer test kit, 89
Sausages, frozen storage of, 106
Scombrotoxic fish poisoning, 38, 39, 43
Screens (windows and doors), 54
SDS (Safety Data Sheet), 56, 57, 151
Seafood. See Fish; Shellfish
Self-audits. See Audits
Self-limiting disease, 30
Service, 125–126
Service animals, 48
Sewage:
    contamination risk, 78
    plumbing, 97
Shellfish. See also Fish
    as food allergen, 40
    identification tags, 38, 101
    purchasing, 101
    storage of, 105, 106, 109
    temperature control during delivery, 102, 103
    toxins in, 37, 38, 39
Shelving, for dry storage, 107
Shiga toxin-producing E. coli, 69, 74
Shigella spp.:
    characteristics of, 23, 28
    employee restrictions and, 69
Shoes, 62
Sinks:
    dishwashing, 92
    food preparation, 117
    hand washing, 66, 91
Skin infections, 63
Slacking, 119, 126
Smoking, 62–63
Solvent cleaners, 86
Soy/soybeans, as food allergen, 40
Spirochetes, 24
Spoilage organisms, 34, 43
Spores, 24, 42, 43
Squash, storage of, 109
SSOPs (Sanitation Standard Operating Procedures), 90
Standards, 141–152
    communicating, 72
    FDA Food Code, 142
    food manager responsibilities for, 152
    food quality, 103
    food safety policy, 141
    for hazardous materials, 151
    inspections for compliance, 143–147
    labeling, 150
    for personal hygiene, 63, 69
    sampling, 148–149

Staphylococcus aureus, 23, 28, 63
State regulations, 4
Stationary phase (bacterial growth), 25, 43
Stock rotation, 104, 107, 110
Stomach flu. See Norovirus
Storage. See Food storage
Submergence method (thawing), 118
Suppliers:
    identifying reputable, 101
    pest prevention and, 54
    screening and approving, 21
    transportation and delivery by, 102–103

T

Tableware, 83
Target levels (HACCP), 134, 138
Tasting food, 64, 120
TCS foods:
    bacterial growth and, 27
    cooling of, 123
    cross-contamination prevention, 117
    daily checks of, 104
    defined, 25–27, 43
    examples of, 114
    food surface contact with, 88
    labeling, 126
    prepared under ROP methods, 110
    reheating, 124
    shelf life of, 105
    storage of, 103, 105
    time/temperature abuse prevention, 114
Temperature:
    ambient, 110
    for bacterial growth, 26
    during delivery, 102–103
    for dry storage, 107
    for frozen foods, 106
    minimum internal, 121–122, 126
    for refrigerated foods, 105
Temperature danger zone:
    bacterial growth and, 26
    defined, 43, 114
    illustration of, 113
    minimizing time in, 117
Test runs, for inspections, 145
Tetrodoxin, 39
Thawing methods, 117–119
Thermal cleaning, 86
Thermistors, 116
Thermocouples, 116
Thermometers:
    for cooking, 115–116, 120
    for refrigerators and freezers, 106
Time, for bacterial growth, 26
Time/temperature abuse, preventing, 114–115, 117
Tolerances (HACCP), 134
Tomatoes, storage of, 110
Toxic (term), 25, 43

Toxic metal poisoning, 35–36, 43
Toxins, 10, 14, 37–39
Toxoplasma gondii (T. gondii), 23, 28, 43
Traditional inspections, 144
Training. See Employee training
Transmission process, 30, 43
Transportation of food. See Delivery
Tree nuts, as food allergen, 40
Trichinella spiralis, 32, 33, 43
Trichinosis, 32, 43
Turkey. See Poultry

U

UL (Underwriters Laboratories), 84
Underwriters Laboratories (UL), 84
U.S. Department of Agriculture (USDA), 5, 6, 142
Use-by date, 106, 110. See also Date labels
Utensils, 82–83, 88, 120

V

Vacuum-packed foods, 102, 110
Validation stage (HACCP), 136, 138
Veal, 108, 121
Vegetables. See Fruits and vegetables
Vegetative state (bacteria), 25, 43
Vehicles (viruses), 29
Ventilation, 80
Vents and ducts, 54
Verbal communication, 72
Verification (HACCP), 131, 136–137, 138
Vibrio parahaemolyticus, 23, 28, 43
Vibrios, 24
Vinegars, and bacterial growth, 27
Viral gastroenteritis, 17, 29, 43. See also Norovirus
Viruses, 29–31, 43
Voluntary closures, 146–147

W

Walk-in refrigerators, 78
Walls, 53, 79
Warewashing, 92, 97
Waste disposal, 52, 117. See also Garbage storage areas
Wastewater, 97
Water activity, 26, 27, 43
Water hardness, 86
Water supplies, 95–96
Wheat, as food allergen, 40
Wild game, 32, 108
Wildlife, as pests, 48, 55
Windows, and pest prevention, 53–54
Wood utensils, 82
Workflow, 78
Written communication, 72

Y

Yeasts, 32, 34, 43